245 $3^{23} 2/22/16 Better world (2000)

read 2/28/16 –

Travels with the Fossil Hunters tells twelve stories of expeditions to remote parts of the world in search of diverse fossils such as dinosaurs and human ancestors. Palaeontologists relate the problems and curiosities they encounter whilst working in extreme field conditions, from the deserts of the Sahara and Yemen to the frozen wastes of Antarctica, from the mountains of India to the forests of Latvia. They tell us what field expeditions are really like: dodging bullets in west Africa or rabid dogs in Pakistan, surviving yak-butter tea in Tibet or raw fish in China. Along the way they also describe the palaeontology and geology of the countries they visit and the scientific reasons for their expeditions. Copiously illustrated with spectacular colour photographs, this fascinating book will appeal to anyone interested in travelling and fossils, amateurs and professionals alike.

Plotting the position of a sauropod dinosaur's
thigh bone before excavation begins: Cretaceous,
Republic of Niger

Travels with the
Fossil Hunters

Edited by

Peter J. Whybrow
The Natural History Museum, London

THE
NATURAL
HISTORY
MUSEUM

CAMBRIDGE
UNIVERSITY PRESS

PUBLISHED BY THE PRESS SYNDICATE OF THE UNIVERSITY OF CAMBRIDGE
The Pitt Building, Trumpington Street, Cambridge, United Kingdom

CAMBRIDGE UNIVERSITY PRESS
The Edinburgh Building, Cambridge CB2 2RU, UK http://www.cup.cam.ac.uk
40 West 20th Street, New York, NY 10011–4211, USA http://www.cup.org
10 Stamford Road, Oakleigh, Melbourne 3166, Australia
Ruizde Alarcón 13, 28014 Madrid, Spain

First published 2000

Printed in the United Kingdom at the University Press, Cambridge

Typeface Proforma 10.5/15.5. System QuarkXPress®

A catalogue record for this book is available from the British Library

Library of Congress Cataloguing in Publication data
Travels with the fossil hunters / edited by Peter Whybrow.
        p.   cm
    ISBN 0 521 66301 6 (hardcover)
      1. Paleontologists – Travel.   2. Paleontology – Field work.
    I. Whybrow, P. J.
    QE714.7.T7   2000
    560′.92′2 – dc21   99–30134 CIP

Travel, in the younger sort, is part of education; in the elder, a part of experience.

Francis Bacon. *Essays – Of Travel*, 1625

Contents

Foreword

Sir David Attenborough, CH, FRS

A tiny glinting facet in an otherwise dull slab of limestone may be enough. Your heart will miss a beat, your pulse speed up and a sparkle comes to your eye. A gentle tap with your hammer. The limestone splits – and there before you, coiled and shining in the rock, lies a sea shell. It is unlike any alive today. It may be 150 million years old. It could be 500 million years old. But one thing is certain – yours is the first human eye to have seen it. That thrill has driven numberless generations of children to search cliffs and quarries, sea-shores and gravel pits, and started them on long journeys of discovery into the past of this planet and the creatures that have lived on it.

For most of us, that flame of enthusiasm has been allowed to dwindle to a hidden glimmer as we grow older and have to earn a living, but it takes very little to make it blaze again. The rekindling is sweet, but it may also be initially painful.

I well remember arriving on the coast of northern Australia and being told by the naturalist with whom I was staying that the beach beyond his house was rich with fossil crabs and crayfish. He showed me some superb examples, peering from their soft limestone pebbles, their shells resplendent with spines and tubercles like the armour of a medieval knight. My mouth watered. The next morning, he took me down to the beach and we walked for several miles to its farthest extremity. Every few paces, he bent down to pick up a claw, a leg, sometimes a complete specimen. I, though I stared with all the concentration I could muster, to my misery and frustration found nothing. And then, as we started on our way back, I found a claw for myself. Thereafter, I almost matched him, carapace for carapace. In short, it may take you a little time to get your eye in. But when you do, the pleasure is just as sweet as it always was.

The writers of the pages that follow have their eyes well and truly in. They are among the lucky ones who, for various reasons that must certainly include both great expertise and abiding passion, have been able to turn their early enthusiasms into

David Attenborough with dinosaur in the
Republic of Niger.

their careers. Not surprisingly, they have made discoveries that make the rest of us deeply envious. As we move round a museum looking at their finds, it may be difficult for us to sense the excitement they felt when they first found these marvellous objects hidden in the rocks. Science, very properly, requires that important fossils should be meticulously prepared to reveal every detail of anatomical evidence. The process may take months, even years and those of us who have tried our hands at it, even in a small way, know that removing grain after grain of rock can give just as much pleasure as the initial discovery. At the end of it, the near-naked fossil lies before our eyes so freed of its matrix that we can deduce just what kind of a creature it had been in life. But what that perfectly prepared specimen can seldom do is to conjure up that excitement, that quickening of the pulse, that the person who found it will have felt.

That, happily, is what the following pages will put right as they take us on expeditions all over the world to find some of the most exciting and surprising fossils that anyone could wish for.

Introduction

Richard Fortey, FRS

Palaeontology is still a profession for individualists. In a world seemingly dominated by investment brokers and computer programmers the palaeontologist still retains a whiff of the adventurer – someone who makes a career out of wandering off the beaten track into deserts or mountains. He (or she) is an explorer of the past, an historian of the time before history began. And, like all explorers, to ply his trade the palaeontologist must travel to remote corners of the world – to places where new discoveries can still be made, and where the rocks have not yet yielded up their fossils.

Some people might find it remarkable that there is still so much to discover in rocks. A little thought will show that it is not surprising at all. Consider what has happened in the past 2000 or 3000 years – not just human events, such as the rise and fall of civilisations, the removal of forests and the extermination of animals – but global shifts like climate changes, the silting up of estuaries and the drowning of fens. Historians struggle with the sheer weight of facts to try to delineate patterns and meaning. Yet palaeontologists have 600 million years as their field of endeavour – or even longer; no less than the history of the Earth and the life that inhabits its surface.

This story must forever remain incompletely understood, for the narrative revealed by fossils is imperfect. How many dinosaurs are still known only from tantalising fragments? How many trilobites remain to be discovered? How many organisms, soft as tissue paper or small as motes, will have nothing robust enough to preserve them as fossils in sediment? Yet even with these limitations the task of unravelling what we can know from what we can find is without end. There are only a few palaeontologists – a tiny fraction of the number of mining engineers or oilmen. This small handful of time explorers must describe the immensity of the story of life from a hint here, a fragment there. They know that their life's work will make only a small inroad into what still lies undiscovered in shale or limestone outcrops, or deep beneath the ground penetrated here and there by boreholes. They know that much of what will be found will be discovered by chance, by the lucky blow of a geological hammer in the right place on the right day of the year. But most of these scientists also know that there are places where you can go where new discoveries are more likely to be made. These may be sites where others have been before and observed promising fragments. Or they may be places where rocks of the right age and kind crop out. Whatever the hint that leads them there, the quest for the past is achieved by means of a scientific expedition. Most of the stories that follow are about such expeditions.

The majority of fossils reveal themselves where geological history reaches the surface in rock outcrops. These outcrops are composed of sedimentary rocks, which were once laid down beneath the sea, or in lakes, or on the flanks of vanished rivers. In their early days the sediments entombed the remains – usually bones, shells or leaves – of the animals and plants that lived in the vicinity. The palaeontologist recovers the fragments, and uses them to piece together his version of history. He can reconstruct the past climate from these humble remains, and describe the type of environment millions of years ago: using extinct land plants and animals he might distinguish between an ancient grassland and a woodland. The most intimate details of daily life – a half-digested meal, or abandoned prey – can be resuscitated from subtle fossil hints and buried fragments. Rocks of the right kind do not often crop out in the middle of towns, or in the domesticated environment of suburban parks. There's always more rock in barren places or in uplands (Gibraltar, where Chris Stringer found evidence of Neanderthals, has long been known as The Rock). So expeditions tend to go to places away from the masses.

The geologist's or palaeontologist's perspective on a country is very different from that of a businesswoman. The latter sees the world through Hilton hotels and international airports; she sits in the front of the plane sipping a complimentary cocktail. The palaeontologist is the character at the back of the aircraft with a battered khaki hat and an awkward camera tripod. On arrival, the palaeontologist eschews the chain hotels, and soon heads off into the wilds with his local colleagues. Who knows? Maybe the businesswoman en route to yet another deal gives this global nomad a brief, envious glance.

Some common threads run through stories of expeditions, for they all, to borrow a discredited phrase from the early 1990s, exemplify the principle of 'back to basics'. There is, in the first place, an obsession with that most basic of all commodities – food. If an army marches on its stomach, a palaeontological expedition digs on it. Mostly, we want large quantities of nourishing tucker. We all have our dietary anecdotes. Andrew Smith clearly tired of rancid yak's butter tea in Tibet, Angela Milner baulked at scoffing live (if singed) goldfish in China. Sand in sandwiches is a hazard recorded by half the authors in this book.

My own special gastronomical phobia is the *bêche der mer* or sea cucumber, which is a particular favourite in the southern provinces of China. Delicacy forbids too detailed a description of these sausage-like animals, which wobble in piles upon the plate. The first problem is that they are remarkably difficult to manipulate with chopsticks. Try to pick them up at one end and they slither away over the table. Try to balance them in the middle and they suddenly teeter off to one side and drop on your lap. Your hosts clearly expect you to relish them, so your pathetic attempts to grapple with the objects have to be mediated by appreciative murmurs of 'so very succulent' – or feeble jokes such as 'this is the one that got away!' When you eventually manage to snag one on your sticks you have to whizz it up to your mouth before it falls off, and then drop it into your mouth in one piece so that it can slither away quickly down your gullet. It is not an invigorating experience. When your duty is done, however, your Chinese hosts will polish off the remainder with gusto.

Then there is accommodation. Most academics run their trips on a shoestring (and that was in the good old days), so the better hotels are out of bounds. Anyway, usually the best fossils are not found near hotels. Instead there are tents and shacks, or sleeping out under the stars. Toilet arrangements become crucial in these circumstances. Several of the stories in this book mention the nocturnal sally with shovel and roll: one does tend to save up these trips for the cover of darkness. Once, when on fieldwork in a very remote part of the Sultanate of Oman, I was engaged on this necessary activity when a camel loomed up unseen out of the dark, and sneezed right behind me. When the party stays in the same place for a number of days there is a problem in finding a fresh site. In Australia the custom is to mark the spot with a stick, but in Arabia there are no sticks, so you tend to see dim shapes anxiously scanning the ground with a flashlight to look for signs of disturbance. It all serves to strengthen the bonding experience.

Sleep can be a problem. Even after a long day's fieldwork you can have difficulty in dropping off. In 1990 I shared a shack with eight Thai geologists; we all slept on thin, palm mattresses laid directly on a concrete floor. Within seconds all my hosts had fallen into deep slumbers, leaving me staring into the dark. When the snores began they were all at a slightly different rhythm and intensity. I lay still in anticipation of the time when, by the inexorable laws of periodicity, all the snores would finally coincide in one, enormous snore – it was like listening to eight gasping metronomes. When it finally happened the effect was stentorian. Then there was no choice but to lie there waiting for it to happen again.

It is surprising that we don't get lost more often. Anyone who has worked in a desert will know how easy it is to mistake one wadi for another. In the western USA many canyons have a kind of generic resemblance. They also have discouraging names like Death's Head Canyon, or Fool's Gulch, and I have tried to find my way through the Confusion Range, all the while avoiding Arsenical Springs and Madman's Flats. Nobody really relishes a flat tire in Death's Head Canyon. After a while one longs for names like Plum Pudding Creek, or Duckdown Lake. It is rather surprising that palaeontologists do not get lost more often in the Gobi or Gaza, and this is largely because of the collaborative nature of research. We work with geologists who know the ground as well as we know the London Underground system in the vicinity of South Kensington. They can discriminate one wadi from another. They know how to drive across the deceptive sabkha so that the Land Rover's axles avoid getting bogged. They know if a passing nomad is a peasant or a revolutionary in disguise. They can fix a broken distributor cap with palm twine – or they know someone who can. In short, they are good to have around, and many of the traveller's tales by my colleagues will record a just debt to their collaborators.

The end of all this travail is a collection, neatly bagged and labelled as to locality and geological age. The collection is carried or freighted back to the museum, where it is unpacked with reverence, and often becomes part of the permanent archive. This is still only the beginning, because the scientific description of the fossil material starts at this juncture, and takes longer – far longer – than its collection. A dinosaur is a complicated animal with very many bones, and these can take years to piece together. There are several examples of specimens collected in a researcher's youth and not published until his old age. One might say that after waiting millions of years in the strata, a mere human lifetime makes little difference (project managers seldom agree). But it has to be said that the formal scientific description of a new species of an extinct animal rarely conveys anything of the excitement of its discovery. This is partly a matter of the lapse of so much time, but mostly it is to do with the conventions of scientific descriptions: impersonal, formal and formulaic, they are mostly as exciting as reading the closing prices in the stock market. There's a good reason for this; they need to be precise and systematic, uncoloured by anything other than technical language, so that they can endure for years without modification, and be understood as readily by a reader in Richmond or Reykjavik. What is lost is all the narrative about the expedition, the sleepless nights, the curious food, the rows among personnel, the lost tracks, the duff hammer blow that wrecked an arm bone – in other words, all the real stuff that goes on in a real fossil hunt.

This book tries to redress the balance. It is a collection of stories by British palaeontologists, working in remote areas of the world, about the bits of science that are usually left out. They convey something of the atmosphere around the fire at night, when anticipation of a notable discovery is mixed with vernacular banter about the cook's socks, or complaints about goat stew again. These accounts should give a feeling for the complicated and often messy business of doing real scientific exploration on the ground. Some even record the failure to find anything – for there is no guarantee of success. Serendipity always has a part to play. Several stories recall skirmishes with danger, and the feeling that we have all experienced while in the field – one of astonishment that we allowed ourselves to get into this fix in the first place. The extraordinary fact remains that major new insights have resulted from such a rag-bag of experiences. At a time when so much discovery seems to be programmed by economic need, and executed by white-coated automata, isn't there something refreshing about stories in which a fortunate puncture, or a hunch fed by rumour and proved by chance, can still influence the way we understand the world?

Chapter one

Across Tibet by jeep, pony and foot

Andrew Smith, FRSE

The staple food in this region is tsampa. The Tibetans make a special cult of tsampa and have many ways of preparing it. We soon got accustomed to it, but never cared much for butter-tea, which is usually made with rancid butter and is generally repugnant to Europeans.

Seven Years in Tibet, Heinrich Harrer, 1952

It was the end of my first day of fieldwork at altitude, over 4000 metres high on the Tibetan plateau. After months of planning, the first geological traverse of the Tibetan plateau by western scientists was finally underway and I was sitting inside a Tibetan herdsman's tent more exhausted than I had ever felt before. With me were Mike Leeder, Lao Yin and Xu Jun Tao. Mike Leeder was the sedimentologist of the expedition, a lecturer at Leeds University and an enthusiast for hill walking, of which we were to get plenty of practice. Lao Yin was a stratigrapher from the Chinese Academy of Sciences in Beijing, a thin-faced energetic man who had been on several Tibetan expeditions previously and was largely responsible for planning where we should go. Xu was my counterpart – a palaeontologist from the Nanjing Institute of Palaeontology. They were to be my companions for the next two months as we gathered information on the sedimentary rocks of the Tibetan Plateau.

We had spent the previous day travelling northwards from Lhasa along winding dirt tracks, being shaken and bounced in the back of a Chinese jeep, eventually reaching a small village by nightfall. This was Pangduo, where we were to be based for the next two days. On one side of the river was a cluster of stone-built Tibetan houses and a small monastery. On the other was a more recent farm complex, consisting of a central courtyard surrounded on three sides by low mud and wattle buildings, where we were to stay.

On arrival we were met by the local cadre and ushered into a small dining hall where, to our surprise, we found a party in full swing. It turned out that a first-aid course for all the Chinese cadres of the region had just ended, and they were now getting down to the serious business of getting drunk. While we sat at a wooden bench eating a bowl of rice and some chicken pieces, groups of two or three men came up and sang lustily around us before passing round a mug of dark brown, muddy-looking beer made from barley. Later mao tai, a rice-based spirit that had a kick like rocket fuel, appeared and toasts were proposed, upon which the small glasses were drained and held upside down over the head.

Eventually, after much toasting, we were shown to our sleeping quarters, a bare room with three simple metal beds with mattresses, and crawled into our sleeping bags. Metal basins were scattered about the floor to catch the drips from the leaking roof. Although tired from the journey and feeling the effects of the alcohol, sleep did not come easily that night because of the rarefied atmosphere.

The next morning was crisp and cold, but it looked set to remain fine for the day. Breakfast had been a bit of a trial: a chipped enamel bowl of rice gruel, a bun of steamed bread with the unexpected texture of putty, a small helping of pickled cabbage with a spicy after-taste, and a bowl of ground peanuts sprinkled with sugar. Because water boils at much lower temperatures at altitude, all cooking had to be done using a pressure cooker in order to heat water sufficiently, hence the strange texture of the bread.

After breakfast we climbed back into the jeep to set off along the last 15 rough kilometres to our planned destination, a valley section to the north. The track out of Pangduo got poorer and poorer and eventually disappeared entirely at the small settlement of Urulung. Unlike the farm commune, this was not a modern fabrication, but a traditional stone built complex with white-washed walls and small, narrow windows. Colourful prayer flags flapped from poles above the flat roofs. All the villagers came out to stare at us, at first staying behind a low stone wall. Later, as they got braver, the children began coming up to us and touching our clothing. Women wore black dresses woven from yak hair, and aprons brightly coloured with red and yellow horizontal stripes. Some wore thick khaki-coloured yak-hide coats with long sleeves that hid their hands.

Here we were to hire ponies that were to carry us up the last five kilometres or so to the outcrop. Mike was rather apprehensive about this, never having ridden before, and he was not reassured when the pony that was being brought for him reared up and threw off its rider before cantering away out of sight up the valley. Fortunately, the second pony that was brought turned out to be much more docile, and we eventually all got mounted and set off. In a very strange clash of technologies I was using satellite photographs to navigate by as we rode along through rock-strewn valleys towards the main cliffs.

We reached a small encampment at the mouth of a steep-sided valley and dismounted. This encampment was used during the summer by Tibetans who had driven their yaks up onto the high plateau to exploit the rather sparse grazing that

Figure 1.1 Mike Leeder being helped onto his
pony at Urulung.

Figure 1.2 Nomad tent north of Urulung where I
had my first cup of rancid yak-butter tea

sprang up once the winter snows had disappeared. There were three huts, each constructed of a circular stone wall about 60 centimetres high and maybe 3 metres in diameter. Above the wall rose a low black tent made out of woven yak hair, suspended by a few ropes and wooden poles. The stone base, although uncemented, was obviously a permanent structure, but the tent and poles were erected and dismantled as the herdsmen moved in search of fresh pasture. We left our ponies here in the care of a small Tibetan boy, about eight years old, dressed in coarse woven woollen tunic and wearing a large fur hat with ear-flaps.

Pleased to get off my pony, I set about the task for which I had come all this way: searching for fossils that would provide information on the age and environment of these rocks. I started working up the ravine, climbing from side to side to examine the rocks exposed on either wall. These turned out to be dirty brown mudstones with occasional sandy or fine gravel levels. Within about 15 minutes I had come across the first layer of rock with recognizable fossils. Splitting open a loose slab of mudstone I found the moulds of fossil brachiopods, marine shellfish about the size of cockles that had thrived here some 290 million years ago. Having found the horizon from which these came I then set about making a collection of the fossils that were to be found, wrapping the best samples in tissue and placing them in cloth bags which were then numbered and put into my rucksack. I also took notes on the succession of beds up the ravine and where precisely the fossils occurred.

Later I found several more levels with similar fossils, but there was something rather unusual about the rocks higher up. Although they were largely composed of fine-grained material (indicative of relatively quiet-water settings), there were, here and there, large pebbles lying within the mudstone. Coarse pebbles embedded in a fine mud matrix is rather an unusual combination to find, because it requires some mechanism for transporting the pebbles into what must have been basically a rather deep and low energy marine environment. The occasional large pebble in mudrocks might have been transported out to sea tangled in the roots of a tree washed out in a flood or landslide, but here the pebbles were too common. Mike suggested that the most probable explanation for these beds involved icebergs. The pebbles could have been rafted far out to sea in icebergs, calved from glaciers on some nearby landmass, eventually dropping to the sea-floor as the icebergs melted. Rare striated pebbles, found by Mike, seemed to bear this interpretation out.

Further up the valley, the rocks changed. Mudstones were replaced by much lighter, cleaner-looking limestones which formed more weather-resistant cliffs. In these limestones I started finding corals, mostly solitary horn-like forms, but later more complex colonies. There were also fusulinid foraminiferans, single-celled

Figure 1.3 The boy who looked after our
horses at Urulung.

organisms the size and shape of buttons, that indicated these rocks were about 30 million years younger than the mudrocks I had been examining below. There must have been a dramatic shift in climate between the deposition of the lowest glacial dropstones and these carbonate platform beds with their coral faunas.

By now I had been clambering and hammering for about five hours, my rucksack was almost full, and every ounce of energy had been drained out of me. Even simple physical tasks at altitude are extremely tiring because of the low oxygen content of air, and by now my breathing had become laboured. Even taking a long drink from my water bottle left me panting for air. I started to descend but was stumbling so much that I could hardly stay on my feet.

Eventually I reached the encampment where Xu Jun Tao was waiting with one of the herdsmen and his wife. They invited us into their tent where it was dark, warm and smokey. Built into the inside of the stone wall was an alcove and two relatively flat platforms draped in yak hides that were both bed and seat. The centre was filled with a stone hearth in which a fire was burning. Since there was no wood of any sort at this altitude, the fuel was dried yak dung. I had noticed that the outside of the walls and many of the nearby stones were plastered with round pats of dung, which I now realized were drying prior to storage. The yak dung created quite a bit of smoke that was swirling around inside the tent, trying to escape from the large circular hole at the apex of the tent and which provided the only light.

A pan was balanced on the fire and from this a muddy brown liquid was poured into a cup and passed to me. I accepted and drank this, too tired to take much in. With a hot drink inside me I started to revive: so much so, that when my cup was refilled and passed to me I started to become rather uncertain as to whether I really wanted to drink its contents. A strong rancid smell was assailing my senses and my stomach seemed to be churning over. However, out of politeness I sipped away at the cup until the contents were all gone. I had survived my first rancid yak-butter tea!

Almost 12 months before, I had been summoned up from my room in The Natural History Museum to the Keeper's office on the fourth floor. Bill Ball, the Head of Palaeontology at the museum, was sitting behind his desk. 'You're a fit young man' he said. 'How would you like to go on an expedition across the Tibetan plateau?' Well, there is not much you can say to a question like that! I must have thought for at least half a second before saying 'When can I go?'

It transpired that The Royal Society had negotiated an agreement with the Chinese Academy of Sciences to undertake a three-month geological traverse across the high Tibetan plateau, from Lhasa in the South to Golmud and the Qidam Basin in the North, a total of some 900 kilometres. Professor Robert Shackleton had been the prime motivator and was putting together the team of specialists that would be needed to document the geology. This was pioneering stuff – the Tibetan plateau was a vast area that had hardly begun to be documented geologically, yet it held the key to understanding the structure of Asia and how the continent had been assembled over time. Chinese geologists had been exploring and systematically mapping the region over the previous two decades, but no western geologists had had the chance to work in the area.

Back 300 million years ago the arrangement of landmasses over the globe was very different from today. Australia, Antarctica, Africa, India and South America were united together to form a single large continent known as Gondwana which lay in the southern hemisphere. In the northern hemisphere, separated by a large ocean called Tethys, was the landmass of Eurasia, comprising Europe, Russia and at least part of what is now China. During the Jurassic, between about 180 and 140 million years ago, Gondwana started to break up, with India separating from Africa about 160 million years ago, and then from Australia and Antarctica a little later, around 135 million years ago. India then rapidly proceeded (rapidly, that is, in geological terms: i.e. at tens of centimetres a year) to drift across the Tethyan ocean, eventually reaching Eurasia approximately 50 million years ago. All the sediment along the northern margin of the Indian subcontinent, and along the southern shelf of Asia was crumpled and crushed as the two continental masses met, and the northern margin of India started to push under the Eurasian margin. Eventually movement ceased, but the huge thickness of crust created as Eurasia overrode the Indian plate was less dense than is usual for rocks at this depth in the Earth, and consequently they started to rise, effectively creating the Himalayas and the Tibetan plateau.

However, there remained plenty of unresolved mysteries about this region as the geological history of the Tibetan plateau was only sketchily understood. The collision of India with Eurasia was just the latest of several such accretionary events, and it was thought that there might be several small continental masses sandwiched between the main Eurasian and Indian landmasses, each with its own geological history. The Chinese had just finished constructing a road across the plateau, from Golmud in the North to Lhasa, offering easy access for the first time to central and northern Tibet. Here was an opportunity to carry out a geological survey along the transect of the new road. Furthermore, the Chinese were keen to develop links with western scientists.

The plan was to spend two months travelling along this route recording the geology on either side. In fact, because of the excellent mapping that had been done already by the Chinese, we were able to pin-point key areas that needed to be studied.

The Royal Society team was formed of ten scientists and a surgeon, Michael Ward. Michael had explained to us when we first met that he had never yet had to operate under field-camp conditions and was looking forward to the challenge. However, he did advise us that we might find it more pleasant to have a full medical check-up prior to leaving. Indeed that was one of the pre-conditions of being accepted.

Each scientist in the team was a specialist in a particular branch of geology, and each was to work side by side with a Chinese colleague. Julian Pearce was a specialist in volcanic rocks and their geochemistry, which proved an essential component for understanding the plate tectonic history of the region. Nigel Harris studied plutonic rocks, large bodies of molten crust that had solidified at depth. His task was to date these radiometrically and to use their geochemistry to provide information on when and at what depth these had cooled and solidified. Doyle Watts was a specialist in geomagnetism. He had the most arduous task of us all, since he had to drill carefully orientated cores of rock. These were collected so that, back in his laboratory in Glasgow, he could measure their remnant geomagnetic signal and thus discover the latitude that they had originally formed in. This meant that he and Lin Jinlu had to man-handle a heavy rock-corer drill, along with a tank of petrol and a large container of water (for cooling the drill bit) up to the outcrop, often some considerable distance away from any vehicle access. Mike Coward, John Dewey, Augusto Gansser, Bill Kidd, Peter Molnar and Robert Shackleton were highly experienced structural geologists, each with their own particular interests and who in their cumulative experience must have worked in almost every corner of the world. Gansser, who at over 70 years old was the most senior member of the trip, was the only person who had been to Tibet previously. Back in the 1930s he had entered Tibet, then a forbidden country for westerners, travelling in with herdsmen and pilgrims. Finally there were Mike Leeder and myself, whose task it was to document and unravel the history of the sedimentary rocks that made up the plateau.

We flew into Lhasa in June 1985 and there met up with our Chinese counterparts. The Chinese had taken complete responsibility for all the logistics for the expedition, and a fleet of four-wheel drive jeeps and three large trucks carrying food and other supplies were waiting for us there, having driven all the way from Chengdu in central China.

The first few days were spent in Lhasa itself, making the final plans and preparations, getting to know one another, and acclimatizing to altitude. Lhasa lies at the foot of a hill dominated by the Potala Palace, within a wide valley and surrounded by jagged mountain peaks. The Potala was the most important monastery in Tibet and had been home to the Dahli Lhama prior to the Chinese invasion of 1951. It is a

Figure 1.4 The Potala Palace, Lhasa, set high on a rock dominating the town.

spectacular building with its enormous and precipitous sloping white walls, high narrow windows and central block of blood-red buildings with their shining golden roofs. It covers the entire summit of the hill, dwarfing the building down below in the town. Early each morning we were woken by music that was being blasted across the town from loud-speakers situated on the Potala.

While in Lhasa we visited the Potala, joining a throng of Tibetans and wending our way up flight after flight of steep and winding white-washed steps. Huge white cloth banners hung down over the main outer entrance, made of bedspread-sized sheets sewn together. Inside it was cool and beautifully decorated. All the woodwork was painted in bright reds, greens and yellows and there were murals everywhere depicting Buddhist gods. Small bells suspended from the roof tinkled constantly in the breeze and the smell of yak-butter candles pervaded the air.

Figure 1.5 Golden bells with wind-chimes on the roof top of the Bhuddist monastery at Lhasa.

Figure 1.6 A family group in traditional Tibetan dress on the circular route around the Jo Khan Temple.

A second major Buddhist shrine, the Jo Khan temple, lies in the heart of Lhasa itself. This is the oldest part of the town, and here the houses are mostly two storeyed with whitewashed walls and rows of squarish windows each with a small canopy of white cloth flashed with red and yellow stripes at the base. Leading from the Jo Khan around the core of the town is a circular thoroughfare that was bustling with life. Family groups walked around the circuit, the women with their black yak-hair dresses and brightly coloured aprons, the men generally in white shirts with their hair braided and wrapped around their heads under battered felt hats. Many of the men carried a short knife with a decorated metal handle in their belt, and I even saw one wearing a snow-leopard skin wrapped over one shoulder. Here and there were monks in their distinctive bright crimson gowns, walking, or sitting by the side reciting prayers. The younger women generally had silver boxes decorated with turquoise hung around their necks and wore an assortment of other jewellery. They were dressed in white cotton shirts and black yak-hair skirts, and had their long black hair in plaits, often reaching well below the waist, and decorated with colourful cotton braids.

The route was crossed by what appeared to be bunting, but which on closer inspection turned out to be hundreds and hundreds of prayer flags strung together. Each flag was a brightly coloured woven piece of cloth the size of a pocket-handkerchief covered in Tibetan scripture. At the two furthest corners from the Jo Khan were six-foot high incense burners into which were periodically fed small bundles of a herb that I did not recognize but which gave off a sweet pungent smell. Prayer wheels, cylindrical brass drums on which scripture was written and which were kept continuously rotating by hand, were everywhere. The two biggest of these, each taller than a person, were situated at the entrance to the Jo Khan and were kept constantly turning by Tibetans entering and leaving the temple.

Outside the Jo Khan were lines of Tibetans alternately standing with hands clasped and then prostrating themselves flat on the stone slabs. We joined the queue to enter, passing a monk with a bucket of yak-butter oil. I decided against taking any, although the Tibetans in front of me dipped their hands in, took a small sip and then smeared the rest into their hair. Inside hundreds upon hundreds of candles burned in front of each Buddha or religious image: the smell of yak-butter was overpowering. I was rather surprised to see that some Tibetans had thermos flasks brightly decorated with flowers and obviously imported from China. These, I quickly realised, were filled with melted yak-butter and were being used to top up candles by the Tibetans as they went around the images. It was a great relief to return to the fresh air outside.

Once the geotraverse was underway properly we quickly fell into a daily routine. We worked in small groups, travelling to different areas according to the types of rock we were interested in and meeting only at breakfast and in the evening after fieldwork for the day was complete. Mike, Lao Yin, Xu and I always worked together and each day our fieldwork took us to different outcrops as we worked our way across the plateau. One day we would be working on 170 million year old red-beds with shallow lake deposits containing freshwater algae and small shrimp-like animals called ostracods, the next day we might be working on 330-million-year-old marine limestones and shales with corals and brachiopods or 270-million-year-old coal seams packed with with large broad-leaved plants. Our team usually travelled in the jeep of

Figure 1.7 A Tibetan street vendor.

Mr Li, a cheery man with an odd assortment of crooked and blackened teeth who smoked constantly. He was very proud of his jeep and spent much of the time when we were out at outcrop polishing and washing the dirt off his vehicle, and he was always careful to wear a pair of white gloves when driving.

It is strange to think that the highest plateau in the world, much of it over 5000 metres above sea-level, is largely made up of rocks that contain marine fossils. In fact, one of our goals was to try to find and date rocks that marked the site of ancient deep-ocean floor. Deep-ocean deposits form a very characteristic and easily recognizable suite of rocks. They include molten lavas, spewed out onto the sea-floor and chilled by contact with sea-water into a characteristic pile of pillow-like balls, dark red clays and black cherts. Organisms with calcareous skeletons are never found in these sorts of rock because at such great water depth calcite is rapidly dissolved. Only the siliceous skeletons of microscopic unicellular animals called radiolaria tend to be preserved in ocean floor rocks. We came across small isolated patches of such rocks, defining narrow linear belts. These marked the suture lines between ancient continents and were all the evidence that remained of once extensive oceans.

The fossils I was collecting were helping to fill in the picture of how Tibet had been pieced together. Firstly, from the kinds of fossils that I was finding I could date the rocks. Every geological period had its own distinctive kinds of animal and plant and so I could tell approximately when the rocks had been formed. Secondly, the fossils also provided information on the environment at the time, whether deep or shallow marine, inshore or continental slope, lacustrine or alluvial. This was useful for reconstructing ancient geographies. Finally, differences in the fauna and flora and in the sedimentary history of different parts of the Tibetan plateau indicated that Tibet was made up of distinct geological regions, each with its own separate history. Tibet was thus made up of several discrete continental blocks that, prior to the collision of India, had moved across Tethys and collided with the Eurasian continent.

Once we had travelled more than about 100 kilometres north of Lhasa there were virtually no signs of native habitation. At one point we passed two Tibetan pilgrims walking towards Lhasa. Their progress was painfully slow – after every four or five steps they would throw themselves prostrate on the ground with arms stretched in front. We passed them in our jeep travelling north one morning, and passed them again when we returned that evening. They were still making their slow way southwards the next morning along the same stretch of road as we moved north to a new camp. I reckoned that at their rate of progress they should probably make Lhasa by the time we finished our geotraverse seven weeks ahead.

Figure 1.8 Saddled Yak at Erdegou. To get to some
of the more inaccessible localities on our
geotraverse yaks were used to transport our
equipment and tents.

The country that we were crossing was not at all what I had been expecting. I had imagined that I would be working through a landscape of towering snow-covered peaks and deep and precipitous valleys. But that, as I came to realize, is Himalayan mountain terrain and we were on the high plateau to the north of the Himalayas in rolling countryside with modest relief at best. We were well above the altitude at which trees could survive, and for the most part a sparse covering of grass or scrub was all that grew. In the more protected valleys, however, small alpine shrubs nestled amongst the gravel providing a bit of welcome colour to the landscape. Relief from the rather barren landscape was provided by the many lakes that are scattered across the plateau. These had a brilliant milky light-blue colour due to the fine rock dust suspended in them.

There was disappointingly little animal life to be seen. Marmots were not uncommon, although we rarely saw them for more than a fleeting second as they ran for cover in their burrows. Similarly deer and wild asses were only seen from a distance. Buzzards and other birds of prey, however, were usually somewhere to be seen circling in the sky, while the occasional bright yellow hoopoe flitted past.

About half-way through the expedition we reached Erdaugou. This was the highest point of the trans-Tibetan highway, almost 5000 metres above sea-level. Up until then the weather had been excellent for us. It had remained reasonably dry and, although often overcast and with a fresh breeze, only became cold at night. But here the weather took a dive and for the first time we had to cope with snow cover over the outcrops. We were staying at an army post, constructed by the Chinese to house a small company of soldiers whose main task seemed to be the repair and upkeep of the road. The conscripts, none of whom were Tibetan, were all small, wiry and extremely young-looking. They were uniformly dressed in thin dark green cotton baggy trousers and Mao-style jackets but wore a motley assortment of shoes, some of which appeared much too flimsy to be any use in this terrain. We came across several bands of these soldiers wielding pick axes and shovels, clearing rock falls from the road or shoring up the crumbling margins along the route.

At Erdaugou we were billetted in one of these small army camps, in a long low room containing 10 beds, a few wooden seats and a central metal stove. We arrived on a particularly bitter evening and there was very little fuel to burn. A small blaze was just starting to cheer us up and we were all huddling round trying to get a bit of warmth, when there was an almighty crack and brief flash. The metal chimney of our stove, as the tallest point on the building, had been directly hit by lightening. Despite the chill in the room, we all kept well clear of the fire after that.

The next morning we woke to find deep snow covering all the buildings.

Figure 1.9 Me inside our accommodation with a yak-dung fire burning merrily.

There was nothing for it but to bide our time in the camp, sorting out our field notes and the rock and fossil specimens. Doyle Watts and I spent a few hours scavenging for scraps of wood for the fire in the snow without much success. We lost three days waiting for the weather to abate.

It was then that I realised I had packed all the wrong sort of gear for this expedition! Never having been on such far-flung fieldwork, I had been careful to take with me spares of just about every piece of field gear that I thought I might need: spare geological hammers large and small, spare chisels of all sorts and sizes, spare compass-clinometer for measuring the dip of beds, spare hand-lenses, pens, mapping case, as well as a huge volume of wrapping paper, padding and cloth bags. This filled a whole rucksack! By contrast, Michael Coward who had worked for several years previously on Himalayan geology and was consequently much more experienced with surviving in remote terrains, also brought a spare rucksack, but he had packed this with a huge round of Parmesan cheese and two enormous German garlic sausages. Doyle Watts had also experienced the rigours of fieldwork in remote parts of the world, having recently returned from a stint in Antarctica, and was also sensibly prepared. He had brought a large jar of Marmite and a packet of coffee bags. The food on the expedition was adequate, but after several weeks of steamed-bread and rice gruel for breakfast, canned fruit for lunch and noodles or rice with meat and diced vegetables in the evening, it was a great spirit-lifter when Doyle made us all a cup of coffee and Mike carefully carved off slivers of cheese and sausage. Great inroads were made into the Parmesan during those three days in Erdaugou.

About seven weeks into the expedition we came to the top of a small pass and saw in the distance a mountain range with snow-covered peaks. This was the Kunluns, a range of peaks which formed the northern rim to the Himalayan plateau. Here we reached glaciers for the first time and there were deeply eroded river valleys cutting through great thicknesses of recently deposited alluvium. As we descended through the mountains the temperature rapidly rose and we found ourselves in an arid, desert-like terrain. Flies for the first time started to become a nuisance, and we were all bitten by mosquitos. We also started encountering local people again: not like the dark and swarthy Tibetans that we first encountered, but men and women more mongolian in appearance. Their tents were different, being small, square and made of white cloth without a stone base, and they were herding camels not yak.

Finally, after eight weeks of continuous fieldwork, the road we were driving along left the foothills of the Kunluns and entered the broad sandy expanse of the Qidam basin. There was even less vegetation here than on the Tibetan plateau, and tall dunes flanked the road. Within four hours we were entering the outskirts of Golmud. On reaching the town's hotel we all made straight for the baths putting a severe strain on the hotel's hot water supply. Field clothes that had been worn almost continuously since leaving Lhasa were unceremoniously dumped into the bin. I had been wearing dungarees in the field, and when I came to put on the pair of trousers I had travelled with to Lhasa I was surprised to find that they did not fit me any more: I needed to find a belt to hold them up! Later that day I found a weighing machine and discovered that I had lost almost 8 kilograms since leaving Britain.

The next few days were taken up with sorting the many hundreds of specimens (amounting to more than 250 kilograms in weight) that had been collected over the trip. These we boxed in wooden crates ready to be sent to Nanjing or London for study. We had covered a huge range of geology in those eight weeks and had learned a great deal about the history of the region. We now knew that Tibet was composed of four distinct continental blocks separated by three suture zones marking

Figure 1.10 (*Opposite*) Looking for fossils on the slopes of the Kunlun Mountain pass.

the sites of ancient oceans. Studying the samples in the laboratory, writing up the results and eventually publishing a detailed geological map of the region was to take up the next two years. Fieldwork in Tibet had made me much fitter (and leaner) than at any other time in my life, and also much less fussy about what I ate. I still, however, draw the line at rancid yak-butter tea!

Chapter two

Fishing – and some dinosaurs – in the Sahara

Alison Longbottom and Angela Milner

L'existence des restes des grands Reptiles fossiles dans les séries continentales sahariennes a été reconnu et signalée par quelques-uns des premiers explorateurs. F. Foreau (1904) et E. Haug (1904) faisaient connaître . . .les restes de Poisssons et de Crocidiles et un fragment de vertèbre de Dinosaurien.

Les Dinosauriens du 'Continental Intercalaire' du Sahara central. Albert de Lapparent, Memoirs of the Geological Society of France, 1960

During our time at The Natural History Museum (NHM) in London we, as specialists on fossil fish (Alison) and fossil reptiles and amphibians (Angela), have experienced numerous discomforts during fieldwork in Europe – deep, glutinous mud in brick pits, battering on exposed sea cliffs by winter storms and minimal, temporary accommodation in bizarre bed-and-breakfast establishments. We had often dreamed about finding fossils in a part of the world where the climate was warm, the weather was dry and there was no wind. Some of our dreams came true at different times during the 1980s when a series of expeditions to the Republics of Mali and Niger were jointly organised by the NHM and the Department of Geology, Kingston Polytechnic (KP, now the University of Kingston) led, respectively, by Cyril Walker and Dick Moody.

The NHM's input into the field work was to obtain evidence of fossil faunas that once inhabited two distinct periods of geological time in western Africa. During the youngest period, the Paleocene and Eocene, some 65–35 million years ago, an inland sea inhabited by strange fish with button-like teeth existed over much of western Africa. The older period, the Cretaceous between 130 and about 65 million years ago, was also interesting because part of Niger was known to be a rich dinosaur locality following discoveries by several French expeditions in the 1960s and early 1970s. Their collections included tantalising fragments of a dinosaur almost identical to the NHM's very own 'Claws' – *Baryonyx walkeri*, the fish-eating theropod discovered in a Surrey brick pit in 1983, and the subject of Angela's research. At the time, out of 50 palaeontologists at the NHM only five were female and we were the first to find out what some of our male colleagues have experienced during previous years in the Sahara.

We describe our fossil-hunting adventures in the Sahara in two parts, starting with Alison's fishing trip to Mali and ending with Angela's dinosaur trail in Niger.

Alison's diary

Although the Saharan fish are more than 60 million years old, the thought of them living in what is now a vast desert is bizarre. My fishing trip, one of the strangest things I have ever undertaken, started in 1981. In Mali the previous year the NHM–KP team had discovered a site with an abundance of different types of fossil fish. Consequently I was asked to go to Mali as the expedition's fish expert to oversee the collection of these intriguing finds. As the only female in the team, most of whom I did not know well, I felt well protected but also vulnerable and apprehensive. I went with mixed feelings but great excitement. The plan was for the party of ten palaeontologists, geologists (who also doubled as occasional mechanics), cooks, navigators and translators to travel overland to Gao, a town almost on the southern, Sahel, side of the Sahara. Cyril and William Lindsay and myself from the NHM, and Dennis Parsons, Tony Buxton, Don Smith and David Rolls from KP travelled in a convoy of one truck, a converted, ancient, ex-Army 'Green Goddess' fire engine and one KP Land Rover. We were to meet up with another KP Land Rover somewhere in Algeria. It all seemed a bit haphazard but I had to assume that the drivers knew what they were doing. Suddenly, the meaning of the words dead-reckoning became very important.

We passed through Europe without much incident, staying at truckers' hotels or, when unavailable, some other, slightly more salubrious place. Our 'Green Goddess' truck with its large white-painted body – a 'White Goddess'? – had been converted into a travelling laboratory-cum-camper filled with cupboards, benches, an oven and a fridge. It was very eye-catching and I thought it almost looked like a Red Cross ambulance, a fact perhaps that helped to smooth our way through Europe's traffic and also made it easier for our Land Rover to follow through busy French and Spanish suburbs.

By the time we reached Algecieras in southern Spain to embark on the sea trip to Morocco the team had settled into a getting-to-know-you routine. The Moroccan route to Algeria had to be used because the truck was too tall to be loaded on the boat at Almeria for the easier, direct route to Algeria. Unfortunately there were 'skirmishes' going on between Morocco and Algeria at the time and the Foreign and Commonwealth Office had advised us not to go through Morocco. Despite us telling them that we had to take this route because of the truck's height and 'could they please sort out the permissions?', we arrived at the Moroccan border only to be denied entry. Suddenly stuck in the no-man's land of Ceuta, a neutral, tax-free town on the northern tip of Morocco, we went to a hotel to 'phone the FCO's 'Our Man in Rabat' to try to get him to speed our entry into Morocco. For the next few days some of us trekked each day to the border to see if we could proceed and each day, of course, were told no.

With no budget for extra nights in a hotel, we decamped to the local car park next to the quayside where we made ourselves surprisingly comfortable and, importantly, were not moved on by the ever-present police. The men slept in the truck or on camp beds by the side of it. As the only woman, I had the luxurious one-room apartment of the Land Rover, in which (I am quite short) I could stretch out easily across the front seats. Each night I would tape blue kitchen-roll paper over the insides of the windows as a 'modesty-curtain' and retire to my little domain. Those four nights in the car park at Ceuta were good training because the front seat of the Land Rover was to be my bed for most of the trip through northern Africa until we finally made camp in Mali.

After five frustrating days, permission was finally granted. Passing through the border we hurried on with enormous relief, but wondered when the next bureaucratic hurdle would materialise. The time I spent at this border was my first encounter with the world of desperation and deprivation that we were to meet several times on our journey. Leaving behind us a long traffic queue of worried, stationary travellers trying to enter Morocco, we by-passed many beggars and peddlers. One woman, I recall, with a baby on her back was trying to sell dolls' legs – not the whole doll, just the legs.

Arrival in Morocco at last! We hastened through the spectacular Atlas Mountains towards the border crossing with Algeria. Despite fears of an encounter with either drug gangs or bandits, we passed through northern Morocco uneventfully, except for one incident. Some youths in a car parked by the roadside attempted to flag us down to 'help' them. But we sped on and for a short while they gave chase, very threateningly, and made throat-cutting motions at us. This reinforced our determination not to stop and, luckily, our driver knew about this clever ruse – he had been this way before! There were no more incidents, and I savoured the wonderful scenery in the Atlas Mountains and its breathtaking sunrises. The border crossing into Algeria at Oijda, where we had the usual two-hour wait at the checkpoint, was also uneventful. Two hours seemed to be the standard at all the borders unless a fuss was created, in which case entry took much longer. I saw a group of Italian youths who had become boisterous and cheeky with the border guards, but they soon quietened when ordered to empty all the contents of their camper onto the roadside. A guard then drove it somewhere for a more thorough search. We left the lads sitting miserably on their belongings, waiting for the return of their transport – some wait no doubt! From this example, we subsequently remained polite, passive and quiet and easily passed through the remaining checkpoints. On the other hand, our easy passage might have come about because most of our store of alcoholic drink, purchased in Spain, was requisitioned by the customs officials who always claimed it was 'for their chief.'

On day 12 we were travelling due south and almost parallelling the Moroccan/Algerian border in the area that the FCO had warned us about. On that memorable night, camped by the roadside, we commented on the unusual amount of traffic passing by, most of it of military; convoys of armoured vehicles, trucks and other war-like hardware. At about 2 a.m. I was awakened to see my kitchen-roll curtains shaking as our leader, Cyril, furiously banged on the Land Rover's window. Even after a week in Africa, he looked rather white and was accompanied by an armed and not very friendly soldier. Soon, we were all awake and ordered to get in our vehicles, follow the soldiers' transport to a small unknown town where we were herded straight to an ominous-looking gate in a barbed-wire topped wall. I thought 'This is it! All my worst fears about this trip have come about, and I and nine men would disappear into an army prison never to be seen again.' At the last minute, though, we turned to one side, skirting the 'prison,' and, further into town, the military brusquely ordered us to park and then departed. It was only when day dawned that we realised that they had left us smack in the middle of the town's rubbish dump. Despite our distant acknowledgement of their apparent concern for our safety, we didn't hang around for long.

We had arranged to rendezvous with the second Land Rover, the rest of the KP team, at a hotel in Adrar, but it was closed. So we went on to Reggane near the start of real desert to load up with supplies, mostly of water and petrol because we had the bulk of our food in our 'White Goddess' – our mobile base camp. The Land Rover's roof rack carried about 20 five-gallon petrol jerry cans (historic stories of the Desert Rats came to mind) and several more stashed in the truck. More fuel than we had bargained for was being used and the problem loomed of being stuck somewhere without any. A very large oil drum was purchased and tied down in the back of the truck and with this full we reckoned we had enough to drive to the next petrol depot, about four days' drive. A piece of plastic piping from the drum out of the back door of the truck helped ventilation and, hopefully, prevented the truck from filling with fumes. None of us smoked but it was still a smelly and scary experience to be in the truck with the drum bouncing around like a potential bomb.

There was only one road on our route and we decided to park beside it until the second Land Rover came by. Luckily we had to wait only one day, which was not bad timing considering we had been travelling for almost two weeks. Although the 'White Goddess' was very visible against the dirty-brown colour of the desert, it still surprises me that we managed to meet and that this was not the haphazard arrangement that I had originally thought. Suddenly, we arrived at the end of the

Figure 2.1 A Saharan desert landscape.

road – literally – and I saw what was, for me, the most surreal sight of my Saharan travels. The tarmac road simply and very abruptly ended, and desert began. I don't know what I was expecting, perhaps a deterioration from tarmac road to a dirt track and then desert, but not this sudden end. It was as if the road contractors had suddenly given up their task. Ahead of us we had five days of real desert travel.

To my surprise there was a recognisable road – the *piste*. Numerous, huge multiaxled trucks ply this route from the Mediterranean, south to countries bordering the Gulf of Guinea, and have formed a road of hard packed but very rutted corrugated sand, so that there was constant juddering throughout all vehicles, loosening nuts, seats, field equipment and our own bones. There was also a constant influx of fine sand into the vehicles that covered everything and everyone. Each evening we had to wipe all our utensils before we could start to prepare our meal but we soon became used to the taste and feel of Saharan sand in all our food.

I had been looking forward to viewing the stars without interference from any artificial light source, and I wasn't disappointed. It is true what other writers have said about this nocturnal scene – you can, almost, reach out and touch the heavens. My other goal was to be able to experience total silence, for at night in a dry open place like the Sahara there are no noises from wind, trees or animals. The silence was indeed total,

Figure 2.2 The 'White Goddess' and Land Rovers at the Tropic of Cancer sign.

but, unfortunately, I never really enjoyed this sensation fully because the constant engine noise from the vehicles' daily juddering left a continuous ringing in my ears for hours afterwards. After we had camped for several weeks I began to realise that we always have sounds inside our heads but we do not usually notice them during the hurly-burly of everyday life. Nevertheless, the effect on my senses that the desert catalysed was an experience that I will never forget. One personal dilemma came about during my nightly trips away from camp with the spade and toilet roll. Our patch of the Sahara was remarkably flat and I had to judge how far away from the camp's lights to go. It had to be far enough not to be seen, but not too far from the light for fear of the surrounding absolute blackness and what it might contain.

The journey through the desert went well although we had a couple of stops to mend punctured tyres. We had taken several sets of bicycle repair outfits, which became very popular as payment to the locals in the towns for their help in repairing our tyres. By day 18 I thought we might have needed them for real as we spied a surreal set of bicycle tyre tracks of fairly recent origin. We followed but never saw the intrepid cyclist; we later learnt that he reached Gao safely.

After five days of heat, dust and filtered, tablet-purified water – we managed without having to resort to the evil-smelling water drawn from a well frequented by camel drivers and their camels – we arrived in Gao. We headed straight for the luxury of the Atlantide Hotel bar, and I downed an iced cola in seconds. Warnings about ice in your drinks in hot countries were completely forgotten. Luckily I didn't suffer any Saharan revenge and later found that the hotel had a good purified water supply. What the hotel didn't have, however, were hot showers. The place was a leftover from the French colonial era and must have been splendid in its heyday. My room was a reasonable size, even had mosquito nets, but the en-suite facilities consisted of a tiled area in the corner of the room and a bucket of cold water. It was only halfway through my 'shower' that I embarrassingly realised that more water could only be had from a tap located in a yard at the back of the hotel!

We spent two nights in Gao enjoying the bustling markets, a trip on the Niger river and visits to the local bars. We also picked up mail at our *poste restante* and it was strangely emotional getting word from families and friends so far away; in the excitement of the journey most feelings of homesickness had been pushed to the back of my mind.

Finally, after more than three weeks of journeying, we reached the site where we were to camp and do our work for the next three weeks. With the inside of the truck being used for the kitchen and storage area, it was a cosy arrangement. The men slept under the stars on camp beds but I had a small tent in which I was very happy because it kept out all the creepy-crawlies that deserts are said to be famous for. One evening, though, when on my way to 'ablutions' I nearly stepped on a large scorpion. My yells brought the whole camp running in eagerness to see the first scorpion so far found. After that, I soon found another spot for my 'ablutions.'

Our work site was full of fossils. We particularly examined the phosphate

Figure 2.3 Upper (left), 8 centimetres long, and lower (right), 5 centimetres long, tooth plates of a pycnodont fish.

beds that had been laid down in a phosphate-rich retreating and shallowing sea. These beds had been eroded by wind and water and the sand surface was now strewn with the dark-brown phosphatic lumps. It was the most fossil-rich site I had ever seen and, when I first saw it, I sat for awhile contemplating the superabundance of treasures for my research. I would walk around flipping over all interesting lumps, my survival knife became a very useful collecting tool (does the Swiss Army know about this use?) and many times I would discover yet another lovely fossil fish. We collected beautiful sets of teeth from a fish called *Pycnodus*. The teeth are pebble-shaped and arranged in rows on the jaw bones. These strange dentitions resemble half a corn-on-the-cob and were used by the fish for grinding its food. We also found hundreds of vertebrae, many fin spines and many skulls of both large and small species of catfish, and several exquisitely preserved skulls of a perch-like fish. Other exciting finds included lungfish teeth and vertebrae of *Paleophis*, a large, extinct sea snake.

This was completely enjoyable and exhilarating work. It was idyllic simply wandering around in the sun collecting fossils in the mornings, having a nice siesta to avoid the hottest part of the day and afterwards collecting more fossils without having to do too much digging. In the evening we would sort through our treasures and pack them safely for the journey home. For most of the time there was just a slight, hazy cloud cover. This reduced the impact of the desert sun to a bearable heat in which we could work during the mornings and late afternoons. Although we had some camp

Figure 2.4 Lungfish tooth plate; scale in centimetres.

lights we settled into a routine of going to bed soon after sundown and getting up just after sunrise to take advantage of the coolest part of the day, something I am almost incapable of achieving when I am at home.

After the KP scientists had plotted the geology of the area where we were collecting, we headed south to Samit, another good fossil locality. This site had a more enclosed area for our camp with some sparse shrubs and trees for shelter. Again, we would go off daily to collect from similar phosphatic deposits. Here we found remains of turtles, including an almost complete shell, about 60 centimetres long, of a very large species, which we set about excavating. This involved sweeping off the sand from the shell and then digging a trench around it. Our activity caused great interest and confusion among the few local nomads who appeared from nowhere to watch us, especially me – a woman in trousers – each day. After trenching, we then covered the shell with a protecting coat of plaster of Paris (the undrinkable well water came into use here). The fossil was then levered out and turned over, more plaster was applied and it was then loaded into the Land Rover, a feat that involved quite a few people because of its size and weight. This plaster lump later caused much interest to the customs officials when we re-entered Spain. They were satisfied only after they had bored a hole into it and let their sniffer dogs have a smell. The dogs were very interested in the 60-million-year-old turtle bones but seemed disappointed that they were not 'prohibited substances.'

Figure 2.5 The large turtle carapace during excavation.

From Samit we visited an older, Cretaceous, site to look for ammonites and dinosaur remains in rocks deposited along a stretch of ancient Saharan seashore, but without much success. We did, however, come across a layer full of sea urchins – echinoids – at a place where they are exposed for what seemed like miles. They cover almost every inch of the surface and look like pure-white, stony pebbles about 5 centimetres in diameter. We couldn't help walking or driving over them, something palaeontologists usually try to avoid. This was the most prolific mass of fossils I had ever seen. We collected only a few because they all seemed to be the same species and, in any case, the previous expedition had already collected several hundreds.

All too soon it was time to start the journey home as we had to be back by Christmas and wanted to avoid the so-called rainy season in Mali. We retraced our inward route and gladly added (our apologies to conservationists) the large, now empty, oil drum to other *piste* markers – sand-polished carcasses of cars, the occasional camel skeleton as well as the ubiquitous oil drums.

We decided to enter Morocco at the southern border crossing with Algeria at the wonderfully named town of Figuig. Algeria was no problem but we could not enter Morocco and, yet again, we were stuck between countries, in a no-man's land a few hundred metres wide between two border controls. Protracted negotiations allowed some of the team to travel in one vehicle to Oujda, several hundred kilometres north, to get the necessary papers for all of us to continue. The remainder spent a very worrisome night sleeping in the truck and the remaining Land Rover, trying to ignore the armed soldiers that constantly patrolled the area around us. The next day the permission-seeking team returned with good news and late that evening we reached Oujda, where we had a hedonistic night in a luxurious hotel with baths and hot water, the first for weeks.

The return journey, as all return journeys seem to be, was very fast. My main memory is of the lunch of sausages, chips and beans I had on the channel ferry, my first taste of British stodge for about ten weeks and I loved it. Food, and toilet facilities, had been an obsession on the trip, especially during the weeks in the desert. In Britain, we were greeted with one of the worst snowstorms for a long time, quite a shock as we had been sunbathing in southern Spain only days before – what a welcome back! Overall, being driven to the Sahara and back and the opportunity to find many fossil treasures was a unique experience for me.

A few years later Angela missed the 'overland experience' when she flew to Niger to join the fourth and largest expedition in the NHM- joint ventures to the southern Sahara, this time to find Cretaceous vertebrates especially dinosaurs.

Angela's diary

In January and February 1988, NHM palaeontologists headed for the desert once more. Most of the KP group were to undertake measuring and dating the sequences of rocks across the Iullumeden Basin in northern Niger and to collect samples for researching the Earth's magnetic field and the chemistry of its crust. Again, Cyril Walker, veteran of previous desert trips to Africa, led the NHM contingent of myself, Peter Whybrow, Andy Currant and photographer, Phil Crabb. Cyril and Andy completed the whole journey overland, a round trip of some 16 000 kilometres (10 000 miles) which took nearly three weeks each way. We were joined by: an ebullient New Zealand zoologist, expert mechanic and truck driver, Graham Wragg; David Ward, a professional veterinary surgeon and amateur fossil shark researcher; Alison Ward, an experienced collector; and Stephen Bankler-Jukes, a professional film director and amateur archaeologist who, as one of our main sponsors, provided us with a reconditioned Bedford MK ex-army truck. This was specially refurbished and fitted out as a camping truck to function as carry-all, camp headquarters, kitchen, wind break and Emergency Ward 10. In addition to the truck, the expedition fleet consisted of three Land Rovers and a Toyota Land Cruiser, loaned to David Ward under Toyota sponsorship. The Bedford replaced the elderly 'Green Goddess' that had been abandoned in Algeria.

The expedition had been planned for more than a year but when the overland party left Dover on 20 December 1987 there was nothing in writing from the Niger authorities although our application for permission to collect had received increasingly positive verbal assurance each time we contacted Niamey. So the expedition set off in ambivalent mood, relieved to be on the way but with the nagging thought that it might be home sooner than planned. As I waved the overlanders off at Dover docks my thoughts drifted back to the newspaper headline after a pre-departure press conference at the NHM – 'Dinosaur hunters ready to risk desert of death.'

It was just as well that I did bring up the rear of the party. A few days before I was due to leave the UK an urgent message was telexed back to London: 'Please bring Bedford steering box.' The truck had suffered the corrugations of the *piste* and its ability to last the return trip was causing concern.

Meanwhile the paperwork and permissions had been sorted out in Niamey and an *autorisation* to collect anywhere in Niger had been issued, anywhere, that is, except Gadoufaoua in the Ténéré Desert, which was precisely where we wanted to work. I had had fond hopes that we might emulate our 'Claws' activities of 1983. But Gadoufaoua had been declared the 'President's Special Reserve,' which meant that it was a conservation area and officially off-limits. As it happened, the area was inaccessible at the time, because of severe sand storms and shifting dunes.

I flew out to Niamey with a large oily package containing the steering box, Stephen Bankler-Jukes, Sir David Attenborough and a five-strong film crew from the BBC Natural History Unit. They were to record the excavation of dinosaurs for a television series about fossils called *Lost Worlds, Vanished Lives*. So it was imperative that the advance party had located suitably picturesque material before the BBC arrived. Fortunately, a combination of an old French Geological Society Memoir and advice from the local bush telegraph, which by this time was well aware of our quest, led them to Izan, an area some 25 kilometres north-east of Ingall.

Ingall is a small town about an hour and a half's drive from Agadez, a dusty treeless outpost, that was once on the main trans-Saharan east–west caravan route. Following reports of dinosaur bones in that area, confirmed by the local Chief of Police, the expedition found the remains of six sauropods within a day of arriving. Graham and Rob Ravenhill from KP met us at Niamey airport with this promising news. Attending to essential bureaucratic matters that ensured the right stamps in our passports for onward travel and shopping for extra supplies and provisions occupied the next day. Then we all piled into the two Land Rovers, together with a huge consignment of filming equipment and several five-litre flagons of Algerian red wine lashed to the roof racks, and set off on the two-day, 720-kilometre journey to Ingall.

Altogether, we found the remains of some 20 sauropods in an area of about two square kilometres; some were well preserved but most had been reduced to heaps of tiny fragments by natural weathering under harsh desert conditions. Some of the bones were preserved in 'log jams' – evidence that dinosaur carcasses had broken up and been deposited by ancient flood water that had swirled the bones into tangled masses. Teeth of theropod dinosaurs were associated with the bones, suggesting that the remains may have been scavenged by carnivores.

The team worked for almost three weeks to excavate more than a hundred large bones, probably belonging to a single individual. In addition, material from other sauropod bone scatters, including teeth, skull and lower jaw fragments were recovered by crawling across the desert on hands and knees. These proved to be the first definitive evidence of camarasaurid-like sauropods from Africa. Camarasaurs, previously known from North America and from very equivocal fragments in southern Africa, are shorter, stockier, short-faced relatives of *Diplodocus*.

Sir David and the BBC crew filmed all stages of the excavation over about eight days so the hardships and horrors of the desert were well and truly captured.

It was indeed hard and horrible! The camp at Izan was set up in the lee of a hill reminiscent of a derelict cinder heap. The surrounding terrain was a flat and

Figure 2.6 Peter Whybrow begins to excavate the femur (thigh bone) of a camarasaurid sauropod. Each femur is 1.65 metres long. A slightly shorter humerus (upper arm bone) lies in the foreground.

Figure 2.7 Three-dimensional jigsaw puzzles: David Ward, Cyril Walker, Angela Milner and Andy Currant test their skills on fragments of sauropod skulls and vertebrae.

featureless plain of red grit, sand and dust, copper-rich and with a sharp acrid smell. It was extremely windy most of the time, although, fortuitously, we had the best weather while the BBC were in camp. We had the job of feeding and watering 26 people during that period – a logistical headache. Except for the flies, eating was not too much a of problem as we had taken a huge range of tinned and dried foods. But a water patrol had to commute to Ingall almost every day to replenish our supply, which had to be chemically treated and filtered before drinking.

The day after the BBC crew departed the wind returned. It stirred every morning with monotonous regularity as the desert surface warmed, blowing up a dust storm that made excavation next to impossible by mid-day. It is not easy to wrap large bones in sheets of tissue paper in a dust storm; neither is it too pleasant when a wet plaster bandage smacks in your face. So we had to stop for a few hours in the middle of

Figure 2.8 Camp headquarters at Izan, complete with 'luxury' patio sitting/dining area and windbreak, aka the Bedford truck. Tents were pitched in a semicircle a few metres out from the social centre.

each day. Our tents provided some relief from the elements, but every day a thick layer of dust, as fine as talcum powder, covered everything inside. Fortunately, the wind always died down in the late afternoon and desert evenings were calm and clear. The second wind of the day started around midnight and some nights the tents were in danger of taking off.

We became resigned to the gastronomic delights of sand with everything although it was not quite bad enough to spoil Cyril's field cooking – it is amazing what can be concocted from a tin of beefburgers. We had hired a Tuareg guide, Ben Hadid Mohammed, who fascinated us with tales of desert life and Tuareg culture around the evening campfire. Mohammed also did some of the cooking and treated us to the occasional traditional North African '*meshwe*' feast – a whole sheep laced with garlic and olive oil barbecued over glowing embers. He also taught us how to convert a five-metre length of cotton into a *seche*, that desert headgear designed as an all-in-one draught excluder, dust filter, sun hat and scarf. We wore our seches every day and usually managed to keep them from unravelling.

Towards the end of the excavation most members of the party suffered from 'Mal d'Izan', an alimentary affliction of varying severity. This experience was not enhanced by having to dash out of one's tent in the pitch dark armed with spade and toilet roll – no time to find the torch – straight into the teeth of a dust storm. The only other major incident on the health front was when the BBC cameraman was thrown off a local riding camel that objected to his large frame on its back. I had ridden the same beast earlier, complete with lethal-looking three-pronged Tuareg saddle, at the special invitation of one of Mohammed's relatives. It would have been churlish to refuse. But I

Figure 2.9 Peter and Angela apply tissue paper followed by hessian strips soaked in plaster of Paris to the big dinosaur humerus; the start of a long process to encase the whole bone in a protective cocoon.

do not wish to do it again, ever. The same peripatetic beast gave us the run around one morning at the camarasaur site by repeatedly trying to walk all over our carefully excavated bones.

The expedition was a great success. Cyril and Andy drove back to Niamey to deposit me at the airport and also to secure the all-important export permit for our

Figure 2.10 Barbecued lamb for dinner. Mohammed keeps an eye on his culinary *meshwe* masterpiece in typical Tuareg dress of *seche* and ankle length robe.

finds; the others began the unenviable task of packing them and running down the camp. As well as the dinosaur bones, a large collection of fish material from several Cretaceous and Paleocene localities was brought back to the Museum, including many species new to science. The most significant fish find was skull material of an early Cretaceous freshwater coelacanth, hitherto known only from scraps. It turned out to be closely similar to coelacanths of the same age from north-east Brazil. This supports the idea that, at that time, Africa and South America were separated only by the narrowest of seas. We also collected amphibian and other reptile fossils, including turtles, crocodiles and the first record of a shellfish-eating mosasaur (a marine relative of the monitor lizards) from that region of Africa.

We went, we saw and we safely returned together with rich additions to the NHM collections and a continuing source of interesting research projects.

Chapter three

Digging the Rock

Chris Stringer

The Rock of Gibraltar has been a landmark for the peoples of the Mediterranean for countless millennia. In classical times it was called 'Mons Calpe,' the northern half of the legendary 'Pillars of Hercules' through which brave seafarers sailed from the known world to face the hazards of the Atlantic Ocean. Around 50 000 years ago, it was also a dominant landmark, but a land-locked one, for this was the time of the last Ice Age when enlarged ice caps at the poles and on high mountain ranges held much of the world's water supply. Seas, including those around Gibraltar were sometimes 50–100 metres lower than present levels.

The landscape then would also have been different, with sand dunes rolling away from the east face of the Rock, down to a river estuary, now drowned beneath the risen Mediterranean. Herds of game would have roamed the coastal plains around Gibraltar – horse, deer, even elephants and rhinoceros – and the Rock itself was home to ibex (wild goat) and vultures. People also lived there 50 000 years ago. They would have looked out from the marvellous vantage point of the Rock to the surrounding lands. But these people were of a different kind – perhaps even of a different species – from any alive today. They were the Neanderthals, and it is a search for more information about these ancient inhabitants of Europe that has led me to make repeated visits to Gibraltar during the past 15 years.

The limestone of the Rock contains many caves, some of them with important fossilised evidence of the Neanderthals, their way of life and, perhaps, their eventual fate. The Rock was one of the first places to yield up evidence of these ancient people, and we now believe it may have been one of their last refuges before extinction.

The accidental discovery of a fossilised human braincase, or cranium, blasted out during work at Forbes' Quarry, below the North Face of the Rock, very nearly placed Gibraltar in the forefront of prehistoric studies 130 years ago. In

Figure 3.1 The Rock of Gibraltar viewed from the north-west. The Neanderthal skulls from Forbes' Quarry and Devil's Tower were both found below the sheer North Face shown here.

September 1864, 16 years after its discovery, the cranium was exhibited at the annual meeting of the British Association for the Advancement of Science in Bath by George Busk, and Hugh Falconer suggested that it be made the type of a new human species *Homo calpicus*, named after Mons Calpe. But this proposal was overtaken by the publication in the same year of the species name *Homo neanderthalensis*, by William King, based on the Neander Valley (Feldhofer) skeleton from Germany, the first time a new species of human had been properly and scientifically proposed. The German skeleton thus got most of the scientific attention. Such famous scientists as Thomas Huxley, Rudolf Virchow and even Charles Darwin commented on the Neander Valley find, but ignored the equally informative Gibraltar specimen. If things had gone differently, we might have talked of 'Gibraltar Woman,' because the Forbes' Quarry individual is judged from its size and shape to have been a relatively small female, rather than 'Neanderthal Man!'

The Forbes' Quarry cranium had to wait nearly 50 years before it got the attention it deserved, from the next generation of scientists, such as William Sollas, Wynfrid Duckworth, Marcellin Boule and Arthur Keith. Although it could not be (and still has not been) accurately dated, comparative studies of its anatomy soon showed that the Gibralter specimen was similar to other western European Neanderthals from sites such as the Neander Valley, Spy (Belgium) and La Chapelle-aux-Saints and La Ferrassie (France). They represented an ancient European population that was often

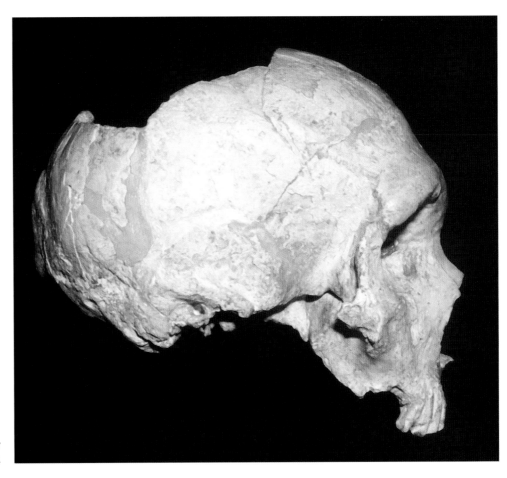

Figure 3.2 The Neanderthal skull from Forbes'
Quarry, found in 1848.

associated with cold-adapted Ice Age animals such as reindeer and mammoth, and
with characteristic Middle Palaeolithic (Middle Old Stone Age) stone tools (also called
Mousterian tools, after the site of Le Moustier in France). These early humans had large
brains, housed in relatively long, broad and low braincases, and long faces dominated
by large noses, surmounted by double-arched brow ridges. The body skeleton, where
preserved, suggests a relatively short, stocky physique, with powerful muscles.

The evolutionary lineage of the Neanderthals of Europe and western Asia
probably separated from that of modern humans (*Homo sapiens*) by 300 000 years ago.
This view, originally based on the fossil evidence alone, has received strong support
from the recovery of ancient DNA from the Neander Valley skeleton itself. Its DNA is
distinct from that of all living people so far sampled, and indicates that the
Neanderthal lineage may have begun its separation from our own as far back as about
600 000 years ago. The best-known ('classic') Neanderthals – those from 100 000 to
35 000 years ago – were adapted to the climatic and physical rigours of life in Ice Age

Europe. Their robust physique helped to conserve heat, and the large cavity in their prominent noses may have acted to warm and moisten inhaled air.

A second significant Neanderthal find was made in Gibraltar in 1926, at the Devil's Tower site, which surrounds a cleft in the North Face limestone, not far east of Forbes' Quarry. This find was excavated systematically, and had associated animal bones, stone tools and charcoal from ancient fires. The fossil remains consist of parts of the upper and lower jaws and braincase of a Neanderthal child. The original assumption that they represented a single child about five years old at death was challenged in 1982 by the suggestion that the bones might represent two children, one aged about three years at death (the temporal bone) and the other about five (the rest of the bones). More extensive studies using a microscope to examine growth lines in the teeth reaffirmed the unity of the Devil's Tower bones, but suggested that the individual might have been about four years old at death.

Most recently, the Devil's Tower remains have been the subject of computerised tomographic (CT) studies using a special x-ray technique. These have revealed new anatomical data, and have produced a three-dimensional reconstruction of the whole skull. This work confirms that the bones belong to a single child but re-emphasises the contrast in size between the two Gibraltar Neanderthal skulls in brain capacity. Although both are close to the modern average of about 1300 millilitres, the volume was probably less than 1200 millilitres in the Forbes' Quarry adult and about 1400 millilitres in the still-immature Devil's Tower individual. Much of this difference may be attributable to previously recognised sex differences in Neanderthal brain sizes, with the adult being female, and the child male. But the skulls are united by one previously unsuspected detail. CT scans also showed that the still-hidden ear bones of these Neanderthals – and of every other Neanderthal studied since – are distinct from those of modern people, and from those studied in the fossil skulls of our probable ancestors.

Although the fossil finds from Forbes' Quarry and Devil's Tower continue to attract scientific interest, and they remain two of the best preserved of Neanderthal skulls, the sites from which they came no longer hold out much promise for further discoveries. The Quarry has a cave in it, but it is now almost empty of sediment, and nobody knows from where the woman's skull was actually blasted. At first glance, Devil's Tower looks more promising, because traces of sediment at the same level as that which yielded the child's skull still cling to the limestone rock face. But the site is in one of the most dangerous locations in Gibraltar, directly beneath the 400-metre sheer north face of the Rock, and subject to regular and potentially lethal rock falls.

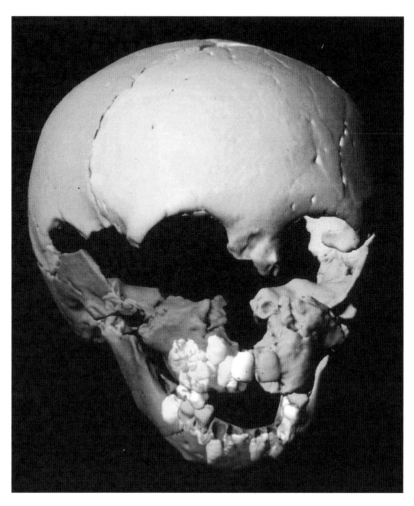

Figure 3.3 A computer-generated reconstruction of the Devil's Tower child, produced from CT scans. Note the images of the unerupted teeth and the ear bones. Reproduced by permission of C. Zollikofer and M. Ponce de Leon.

So in 1984, when I began to examine potential sites for excavation in Gibraltar, I looked beyond the obvious locations of Forbes' Quarry and Devil's Tower, and especially focused on a series of caves near the sea on 'Governor's Beach', on the south-east side of the Rock. Although Captain A. Gorham is credited with the discovery in 1907 of one of these caves, subsequently named after him, they were not systematically investigated until after 1945, when access was still only possible by sea from below, or by ropes from above. Today, the beach mainly consists of fine limestone blast debris from military tunnelling operations, but there are also rocky remnants of a more ancient cemented beach that accumulated during the 'Last Interglacial,' which was the last period of time when the Earth's climate and sea level resembled the present day's. Our present interglacial, called the Holocene, has spanned about 10 000 years, and the previous one, marked by the cemented Gibraltar beach, lasted from

Figure 3.4 The caves on Governor's Beach. The main ones are (from the left) Bennett's, Gorham's, and the double opening of Vanguard and Boat Hoist.

about 130 000 to 115 000 years ago. It was followed by the unstable and predominantly cold climates of the last Ice Age.

The stretch of coastline around Governor's Beach is under military control, which has helped to protect the caves from casual visitors. On the other hand, the area has been used for military training, which means the foreshore is littered with potentially dangerous military debris, and the cave deposits have bullets scattered on them, or even embedded *in* them! There are four main caves in a row facing the sea – from the north, these are Boat Hoist, Vanguard, Gorham's and Bennett's Caves. When I saw them in 1984, Boat Hoist and Bennett's had long been emptied of most of their ancient sediments by marine erosion, a process that has continued throughout the Holocene, because of the high sea level. Nevertheless, the surviving deposits that cling to the cave walls bear witness to occupation by the Neanderthals – their stone tools can still be seen poking from the sediments in places. Vanguard Cave was a pristine site that seemed to be entirely filled with sand, while Gorham's, the largest of these caves, displayed a mixture of disturbed and untouched deposits.

A geologist serving in the British Army, G. B. Alexander, conducted excavations at the back of Gorham's Cave after World War II and, although his work was never published, he found that the highest cave deposits at the back of the cave contain valuable historic and prehistoric material. The cave had obviously been a sacred site for the sailors of Phoenicia, Carthage and Rome for they had regularly visited it to deposit offerings to their Gods, presumably in the hope of securing a safe passage through the Strait of Gibraltar. The dark earthy deposits that Alexander excavated were rich in pottery, coins, jewellery and scarabs (tiny, engraved replicas of the sacred Egyptian dung beetle). The scarabs were worn on rings or pendants, and were used as amulets, or to mark wax seals on ancient documents. The deposits extended for at least a couple of metres depth, and reached as far back as the Neolithic – the New Stone Age period, more than 5000 years ago. Below these layers were much more ancient levels, extending beyond the Holocene and into the Pleistocene, the name given to the geological epoch that lasted from about 1.6 million to 10 000 years ago. It was these levels that contained the stone tool evidence that Neanderthals once lived in the region and in the caves themselves, and it was the excavations of the British archaeologist John d'Arcy Waechter that first revealed them properly.

Waechter supervised a largely untrained team who removed significant portions of the Gorham's Cave sequence from 1951 to 1954 but, in doing so, they established that the cave contained an immense record of human occupation, spanning much of the past 100 000 years. He recorded that the ancient raised beach, which we would now date to about 120 000 years, lay at the base of Gorham's Cave, and hence all the archaeological levels must be younger than that age. Above the base, there was at least 10 metres of evidence of Neanderthal occupation, marked by Middle Palaeolithic tools, and then a couple of metres of deposits containing Upper Palaeolithic tools, which elsewhere in Europe are characteristic of occupation by modern humans, after about 40 000 years ago. Capping the whole sequence were the Holocene levels of the past few thousand years.

Waechter also reported the presence of ancient hearths at various levels in the cave, especially in the Upper Palaeolithic, and animal bones throughout the sequence, dominated by the remains of ibex, rabbit and many species of bird. Unfortunately, many aspects of Waechter's excavations were never properly published, and much of the material he recovered has since disappeared.

Nevertheless, Gorham's Cave did indeed look a most promising site for further excavation, particularly because Waechter had obviously – and

Figure 3.5 Gorham's Cave under excavation
in 1995.

fortunately – been unable to excavate out all the sediments. So in 1989, I organised a short exploratory excavation of the surviving cave deposits in collaboration with the British Museum. Sand blowing up into the cave for 35 years since Waechter had left, and repeated recreational visits by soldiers, had obscured the sequence. But our three weeks of work showed that significant parts of both the Neanderthal (Middle Palaeolithic) and early modern human (Upper Palaeolithic) levels still survived. We were also able to recover some useful bone and stone finds from Waechter's spoil heap in front of the cave, because his team apparently did not screen (sieve) the deposits for small finds before dumping them.

We were helped in the logistics of our excavations by various arms of the British military forces in Gibraltar, including the provision of some soldiers to help us transport all our equipment down the 360 steps that had been constructed to the beach. We were visited by the Governor of Gibraltar, and, as the visit coincided with his birthday, he (and we) were treated on site to surprise bottles of champagne! Military help even extended to the offer of taking all our heavy equipment back off site on amphibious landing craft. This would have saved us a lot of time and labour; it was indeed unfortunate, therefore, that we kept all our heaviest and most awkward equipment on site until the last day, only to find that adverse sea conditions prevented the craft from landing, thus necessitating some back-breaking repeat trips up and down the steps for an already weary team.

Various difficulties prevented our return to Gibraltar until 1994, by which time there was a new Governor, the Gibraltar Museum (with whom we had been collaborating) had a new director (Clive Finlayson) and a new cave was available for excavation. This cave had first appeared in 1974, and lay high up on the east side of the Rock. Much of this part of Gibraltar was covered by a huge natural slope of rock debris and sand, which had accumulated during the Pleistocene. This slope was utilised by Victorian engineers, who began a process of covering it with iron sheeting, down which rain water could run and collect, thus ensuring Gibraltar of a freshwater supply in times of siege. But by 1975, the 'water catchments' were in disuse and decay because Gibraltar had developed new sources of fresh water, and a Scandinavian company began to strip off the top sheets of metal in order to quarry the underlying sand and rock to use down at sea level for land reclamation. As they did so, a small cave opening appeared, which became known as Ibex Cave, because one of the first objects found in it was a well-preserved ibex skull. In the sand infill of the cave, several stone tools were found, and when I was able to borrow them and show them to archaeologists in London, they were recognised as almost certainly the handiwork of Neanderthals.

Figure 3.6 Andy Currant begins the long morning
climb of nearly 800 steps up to Ibex Cave.

Because the cave had hardly been touched, the team decided to carry out a trial excavation. Gorham's Cave had given us logistical problems in 1989, because of the 360 steps, but those were nothing compared with the access difficulties posed by working at Ibex. Although the Scandinavian company was able to drop its commercial excavation equipment by helicopter, assembling diggers and bulldozers on site, we had to carry our equipment up 780 steps to the 260-metre high cave. Not only that, to gain access to the bottom of the steps, we had first to walk for about 800 metres through a narrow tunnel that ran straight through the Rock from west to east. Even worse, that tunnel, which was used by Gibraltar's main water company, was locked at 4 p.m., which meant our return journey had to take us down about 1000 steps to the opposite side of the Rock from where we began. In addition, our hopes of wet-sieving to concentrate the sediments on site were dashed because of a lack of water, necessitating the carrying of many full bags of cave deposit down all those steps.

Ibex Cave produced some interesting evidence of denning by foxes and wolves, but the phase of Neanderthal occupation, which we dated at around 40 000 years ago, seems to have been relatively brief, and this, combined with the severe logistical problems of the site, persuaded us to return to Governor's Beach and Gorham's Cave in 1995.

By the mid-1990s, there had been some significant developments regarding the Neanderthals. Excavations at a Neanderthal site near Malaga in Spain, called Zafarraya Cave, had not only revealed new fossil evidence of the Neanderthals, their stone tools and food debris, but indicated that they had lingered on in this region much later than had previously been suspected. Evidence from France suggested that the Neanderthals had disappeared from western Europe by about 32 000 years ago, dated by the radiocarbon method. But the same dating method suggested that Neanderthals occupied Zafarraya Cave until about 27 000 years ago. This meant that the Neanderthals seemingly survived in southern Spain long after they had disappeared further north, and even longer after the appearance of the first modern people in western Europe, some 40 000 years ago. Further dates from other sites in southern Spain and Portugal appeared to support this scenario of Neanderthal late survival.

The caves on Governor's Beach looked ideal to put these new ideas to the test. They certainly contained Neanderthal camping sites, and Gorham's, at least, had subsequent occupation by early modern people. Our immediate objectives, if we could raise the funding for three seasons of work, were to attempt to date the transition between Neanderthal and modern human occupation, and to compare the way the two different peoples had lived in the caves. As well as our main excavations in Gorham's

Figure 3.7 The excavation team at Vanguard
Cave, 1997.

Cave, we also hoped to investigate what lay under the undisturbed sands of Vanguard.
Excavating the caves properly, for the first time, meant organising a much larger non-
Museum team than in previous field seasons – at times up to 30 people were working
in one or another cave. As well as the various specialists in the types of finds we
expected to make – for example, mammal bones, birds, stone tools – we also needed
assistants to process material both in Gibraltar and London, a full-time site
manager/driver and a specialist to run the sieving operations. Experienced excavators
were recruited to expand the team, and experts in sedimentology, taphonomy (the
study of the processes that accumulate and preserve bones and other finds) and dating
joined us on site. By the time of our 1997 season, we had the use of a powerful generator
to run six lights, with cabling running several hundred metres between the two caves, a
petrol-driven pump to provide high-pressure sea water for the sieves, a military hoist to

carry heavy gear up and down the cliff, and a system of ropes and pulleys capable of transporting sediment bags from the excavations down to the sieving area. Communication, and safety, were greatly enhanced by the use of mobile phones and long-distance walkie-talkies. We were joined by Gibraltar Museum staff and members of their families, local volunteers, who willingly worked through their holidays for us, and a group of Spanish archaeologists, who made a special study of the more recent sediments at the back of Gorham's Cave.

At Gorham's Cave, we were quickly able to uncover the areas that had been cleaned previously and soon got to work in the Middle Palaeolithic levels, but, at Vanguard, we had no idea what to expect. We hoped that under the huge sand slope was a sequence matching that at Gorham's, but, in 1995, there were only hints that this was so. A series of steps were cut into the sand, but most seemed to show yet more sand, with only sparse evidence of bones and stone tools. The 1995 season at Gorham's, by contrast, looked very promising, and we were able to excavate in one of the uppermost Middle Palaeolithic levels where Waechter must also have excavated. There were rich collections of stone tools, bones, burnt nuts, seeds and charcoal, mapping areas where the Neanderthals had built fires and processed meat and vegetable foods. A radiocarbon date for some of the charcoal came out at close to the limits of the method – about 45 000 years – and we planned to work up from that level until we reached the oldest Upper Palaeolithic layers. The following winter was, however, one of the wettest on record in the region, causing extensive erosion in Gorham's Cave. This led to massive amounts of the precious unexcavated Upper Palaeolithic levels collapsing onto the Middle Palaeolithic sediments we had been digging, which meant we had to modify our plans and areas of excavation for 1996.

On only the second day of the 1996 excavations at Vanguard, we had a major shock. A new trench was opened up at the base of the sand slope, and some human foot bones were discovered not far below the surface. But they were not those of the hoped-for Neanderthal burial – they looked distinctly unfossilised, and fresh. To our disquiet, it seemed they might be very recent indeed, and we therefore telephoned the Gibraltar police to report what we had found. Soon, the site was swarming with police, and, to our horror, our excavations were suddenly out of bounds, on the other side of newly installed 'Keep Out! Scene of Crime' barriers. Discussions with a police forensic expert in the United Kingdom luckily improved the situation, however, and we were told that we would be able to continue our excavations the next day, provided we also formed a small team to concentrate on carefully exposing the newly discovered body. We were to screen the surrounding sand for any surviving evidence, and we were to leave the skeleton in the ground for inspection by the UK forensic expert, who was

Figure 3.8 Vanguard Cave at the beginning of
excavations in 1995.

to fly out at the earliest opportunity. Two days later, the skeleton had been carefully uncovered and, from the form of the bones, was clearly that of a teenager, probably a girl. Long dark-brown hair still rather gruesomely clung to the skull, but there was no evidence of clothing or jewellery. The burial had clearly been carried out with some care, and the body had been laid face up, with arms by the side and legs extended. The teeth in the skull were in good condition, with no evidence of dental treatment, and this led the police to surmise that this body was not that of a local, but of someone who had been brought in by a small boat for burial – perhaps even that of an illegal immigrant from North Africa. We provided the police with as much information as we could regarding our unfortunate discovery, but the teenager concerned remained unidentified at the time of the inquest, and an open verdict on the circumstances of the burial was returned.

With this little drama behind us, we were able to conclude our excavations at Vanguard in 1996 with a clear demonstration of the potential of the cave. At least two rich Middle Palaeolithic levels had began to show up, separated by deep stretches of intervening sand, and a small alcove in the north of the site showed intriguing evidence of occupation by hyaenas, including their own bones, their food debris, and their coprolites (fossil excrement).

By 1997, we had obtained a series of radiocarbon dates from charcoal and fossil bones for occupation levels at both caves. They showed that Vanguard, the smaller of the two sites, must have been virtually full of sand and sediment by about 40 000 years ago, and thus was likely to contain significant evidence only of Neanderthal occupation. At Gorham's, however, there were several levels of Upper Palaeolithic material, which we could date to between 26 000 and 30 000 years ago, and these showed that the arrival of modern people in Gibraltar must have occurred while Neanderthals did indeed still live in the more mountainous interior of Iberia. We also had dates by radiocarbon and other techniques on the lower parts of the sequences showing that Neanderthal occupation of Vanguard extended back at least 50 000 years, whereas that of Gorham's went back at least 100 000 years. A particularly remarkable find from the upper levels of Vanguard was the first unequivocal evidence that Neanderthals utilised marine food resources – a topic that has been a source of debate for many years. There we uncovered a discrete Middle Palaeolithic layer consisting of mussel shells of a large and consistent size, mixed in with ash and stone tools, some of which showed edge damage that might have come from opening or scraping out the shellfish. This gave us a completely new view of Neanderthal capabilities. Although we had previously found limpet shells close to ash and charcoal at both sites, these were never in a clear-enough context to indicate whether this was anything more than a

Figure 3.9 A Neanderthal fireplace under excavation at Vanguard Cave, 1996.

purely fortuitous association. Now we could show that the Neanderthals were not ignorant of the benefits of the many coastal food resources lying close to their cave shelters in Gibraltar.

Yet the manner in which the Neanderthals used their caves in Gibraltar did seem to contrast with the way that the succeeding modern people occupied them. Although their diets seemed similar to judge from the ibex, deer, rabbit and bird bones in their respective occupation levels, the fireplaces preserved from the Middle Palaeolithic of both Gorham's and Vanguard seemed to be only that – a place where one or a few Neanderthals dug a hole in the sand and briefly lit a fire. After a short time, perhaps one night, they moved on. But in the Upper Palaeolithic levels at Gorham's, the early modern people seemed to light their fires in the same places over many years, and there was evidence of beach cobbles being used to line the hearths. Their settlements seemed longer-term, more intensive, and perhaps consisted of larger groups.

How often the last Neanderthals in the Gibraltar region encountered the first modern people there, and the manner of their interactions, is something we can

Figure 3.10 A collection of Neanderthal stone tools produced from different raw materials, from Gorham's Cave. Ruler = 14 cm.

only guess at. There may have been violent encounters, they may have avoided each other, or they may have made relatively peaceful contact. From the Gibraltar and southern Iberian evidence, it seems that the Neanderthals did not change their technology when the new people arrived – unlike their counterparts further north, they maintained their ancient stone-tool-making traditions essentially unchanged until they disappeared.

The disappearance of the Neanderthals from Europe by about 30 000 years ago (and probably less in southern Iberia) seems at first sight to neatly coincide with the arrival of modern humans – the 'Cro-Magnons' (named after a French cave site containing their fossil remains and Upper Palaeolithic artefacts, discovered in 1868). But we now know from evidence in both Europe and western Asia that the replacement was anything but instantaneous – it took many thousands of years. So were the Neanderthals gradually replaced following competition from better-adapted Cro-Magnon hunter-gatherers? Similar replacements apparently took place in Asia and Indonesia as the local counterparts of the Neanderthals encountered dispersing modern humans, and superior technology and organisation must surely have played a part.

On the basis of their physique and facial form, it seems likely that Cro-Magnons ultimately derived from Africa, and were part of a radiation of modern humans from that continent, beginning about 100 000 years ago. The impetus behind the modern human dispersal was presumably increasing population sizes but the reasons for our evolutionary success are still uncertain. It was obviously not based on physical superiority (given the strength of the Neanderthal skeleton), but must have been due to such things as language, social organisation and planning. But the world's unstable climate may also have played a part.

Recently, it has become apparent from long records of past temperatures preserved in ice caps, ancient lake beds and on the floors of the oceans that the period when modern humans took over from people like the Neanderthals was a particularly volatile time for climatic change. Every few thousand years, the world's climate switched repeatedly from relatively warm to very cold conditions, and then back again, and many changes may have happened in less than 50 years. The resultant rapid environmental changes would have put tremendous pressures on human hunter-gatherers, who had to live off the land and its resources. In such times, it would have been the most resourceful and adaptable people who survived, and perhaps this is where modern people had the edge over the Neanderthals.

The behavioural evidence of some of the last Neanderthals and their modern human successors that I and my colleagues are retrieving from the Gibraltar Caves will help to give us a measure of how similar or how different these peoples really were in their capabilities, and may throw light on the reasons why we survived and they became extinct.

Chapter four

A summer in Latvia

Per Erik Ahlberg

Half my closet walls are covered with the peculiar fossils of the Lower Old Red Sandstone; and certainly a stranger assemblage of forms have rarely been grouped together;– creatures whose very type is lost, fantastic and uncouth, and which puzzle the naturalist to assign them even their class; boat-like animals, furnished with oars and a rudder; – fish plated over, like the tortoise, above and below, with a strong armour of bone, and furnished with but one solitary rudder-like fin; other fish less equivocal in their form, but with the membranes of their fins thickly covered with scales;– creatures bristling over with thorns; others glistening in an enamelled coat, as if beautifully japanned . . . All the forms testify of a remote antiquity – of a period whose 'fashions have passed away'.

The Old Red Sandstone,
Hugh Miller, 1841

The cold wakes me at 3 a.m., just like the night before. Even though my sleeping bag is buried under a heap of spare clothes, I can feel the heat seeping out through its thin and lumpy lining. The air mattress seems to have deflated as well. There's no chance of sleep now, not until the sun rises above the forest edge in about an hour and starts warming the tent canvas; I might as well take a walk. It is still dark in the tent, but with a bit of effort I bundle on my coat and boots and stumble outside. Dawn is just breaking over our meadow, with its scatter of green tents and old-fashioned cars in the waist-high grass. Nobody else is stirring. The rhythmic, whirring song of bushcrickets rises out of the grass all around. As always after a clear night, the dawn is bitterly cold and there is a heavy dew. Shivering, I decide to head back to the tent again and take my chances with the sleeping bag. At moments like this the attractions of fieldwork seem less than compelling. Why am I here, in a meadow in Latvia, when I could be sleeping comfortably in my own bed in my own house?

The answer lies under my feet. There is sand in my boots, fine greenish Devonian sand. Come to think of it, there's sand in my tent, my hair and my toothbrush as well, but you learn to live with these things. Around 365 million years ago there was a coast here, facing south onto an equatorial ocean. Sediment-laden rivers flowed down from the northern hinterlands, building out their deltas into the sea, gradually mantling the sea floor in a vast blanket of sand, silt and clay. Today the sea and the rivers are long gone, but their sediments remain under the forests and farmlands of the Baltic states. Buried within them lie the fossil bones of animals that once swam through those ancient waters. They provide a glimpse of a crucial step in our own evolution: the move onto land.

Life began in the sea. This was where the first single-celled organisms arose and where, much later, the first complex multi-celled animals and plants evolved.

From the Cambrian period, about 550 million years ago, we have an abundant fossil record of many different marine animals, including early representatives of living groups such as molluscs and echinoderms. Our own group, the vertebrates or backboned animals, first appeared in the form of primitive jawless fishes towards the end of the Cambrian. During the next 100 million years or so vertebrates diversified mightily, evolving jaws and radiating into a plethora of body forms and lifestyles. Yet, while all this was happening in the sea, the land remained largely barren, enlivened by no more than crusts of lichen and algal scum on damp surfaces. Eventually, and perhaps inevitably, life found its way onto land. Algae evolved into plants with a waxy outer covering, which could withstand the drying effects of air. Arthropods such as centipedes and millipedes emerged from the water to forage through the newly formed soil and detritus among the roots of these plants. Once the land ecosystem had been established, it developed quite rapidly. By the middle of the Devonian period, about 380 million years ago, the first forests had appeared. Not long afterwards, vertebrates began to crawl onto the land.

Together with a group of colleagues around the world, I have spent the last ten years trying to track down the origin of the land vertebrates or 'tetrapods.' We didn't have to start from square one; the origin of tetrapods has been a hotly debated research topic for more than a century. A few points were already firmly established. Firstly, that the tetrapods had evolved from lobe-finned fishes, a group which has a long fossil history but also includes the living lungfishes and the coelacanth *Latimeria*. These fishes have paired fins with a unique, limb-like internal skeleton. Secondly, that the origin of tetrapods was a single event, rather than a series of separate and parallel originations. All living and fossil tetrapods (that is all amphibians, reptiles, birds and mammals, including ourselves) share a large number of characteristics which show that we are a 'natural group' descended from a single common ancestor. Thirdly, that the earliest fossil tetrapods came from Late Devonian rocks, about 365 million years old, suggesting that tetrapods originated somewhat earlier during the Devonian period. But there were still huge gaps in our knowledge.

A major part of the problem was that so few really early tetrapods were known. The only genus known from almost complete skeletons was *Ichthyostega*; a vaguely crocodile-like creature, about a metre in length, which was discovered in Devonian rocks in Greenland in the 1930s. *Ichthyostega* had some very primitive and fish-like features, such as a small tail fin, lateral lines on the head (these are sensory organs which detect vibrations in the water) and a paddle-shaped hind foot with seven toes. However, it also had some strange and unexpected features, like very broad overlapping ribs. We needed to know whether these features were typical of Devonian

Handwritten margin notes:

Richard Dawkins, p168 (greatest show)

fish
↑

eusthenopteron (1881)

pandericthys   (amphibian like fish)

Tiktaalik (Shubin)

acanthostega   (fish like amphibian)

ichthyostega 1932   tulerpeton

elginerpeton   (tetrapod)

↓

tetrapod

---

tetrapods in general, or whether they were simply unique specializations of *Ichthyostega*. But in the mid-1980s so few Devonian tetrapods were known that these questions simply couldn't be answered. The hunt was on to find more fossils.

Three notable successes were scored quite quickly. In 1984 a Russian scientist, Oleg Lebedev from the Palaeontological Institute in Moscow, discovered an incomplete but beautifully preserved skeleton of a Devonian tetrapod at a site near Tula in central Russia. *Tulerpeton*, as it came to be known, was smaller than *Ichthyostega* (about 50 centimetres in length) and seemed much less primitive although it was about the same age. In 1987, a joint British–Danish expedition set off for the east coast of Greenland in search of *Acanthostega*, a shadowy genus known only from one incomplete skull. I was one of the participants, having been invited along by Jenny Clack, my PhD supervisor at the University Museum of Zoology in Cambridge. Six weeks clambering over scree slopes and rocky plateaus produced extraordinary results: at least half a dozen more or less complete skeletons of *Acanthostega*. This animal seemed more fish-like than *Ichthyostega*, with a larger tail fin, gills behind the head, and eight toes on both forefeet and hindfeet.

Then, in 1991, I discovered that museum collections of fossil fish bones from Scat Craig, a Devonian site in Scotland, also contained fragments of an unrecognised tetrapod. Now named *Elginerpeton*, this holds the distinction of being the oldest tetrapod so far discovered. It was a large creature, maybe 1.5 metres or so, but the remains are very fragmentary. All of a sudden, Devonian tetrapods were starting to look like quite a diverse group of animals!

While all this was going on, Oleg Lebedev decided to visit his colleague Ervīns Lukševičs in Latvia (then still under Soviet rule). Together they organised a summer excavation at Pavāri, a recently discovered late Devonian locality on the banks of the Ciecere River. The dig was a great success, yielding numerous bones of a curious little armoured fish called *Bothriolepis* as well as remains of lobe-finned fishes. But, puzzlingly, there were also long slender jaws, studded with sharp teeth, which were less easy to identify. Similar remains had been attributed to lobe-fins in the past, but Oleg and Ervīns were not convinced. I visited Latvia for the first time in 1991, together with Oleg Lebedev, and was straightaway taken to see Ervīns Lukševičs at the Latvian Museum of Natural History (LMNH) in Riga. With an air of suppressed excitement they brought out the strange jaws from Pavāri and asked me what I thought. I was stunned. The jaws were almost identical to those of *Acanthostega* and *Elginerpeton*, quite different from any lobe-fin. We had a new Devonian tetrapod on our hands.

Oleg and Ervīns had already made arrangements for a field trip to the site, so we spent the next week camping by the Ciecere with a crew of staff and volunteers

Figure 4.1 The Old Town of Riga.

from the Latvian Museum. We returned to Riga in triumph, our vehicles loaded down with specimens – and discovered that Michail Gorbachev had just been deposed in a military coup. That afternoon in Ervīns office, I tried to steady my nerves and concentrate on my work, while helicopter gunships clattered overhead, Soviet troops moved to take control of the television centre, and Oleg was down at the railway station frantically trying to get us two return tickets to Moscow. We left Riga on the last train that evening, not knowing whether we would ever see Ervīns or the fossils again.

By the time I next set foot in Riga, in the summer of 1992, the coup leaders were in jail, Latvia was a free country and the Soviet Union had ceased to exist. Our research had moved on, too; we had decided to name our animal *Ventastega*, after the Venta River into which the Ciecere flows, and were making good progress with our study of the fossils. But although we had excellent material from the lower jaw and cheek bones, and parts of the shoulder girdle and pelvis, it was clear that large parts of the skeleton were still missing. We started to plan another excavation.

Scroll forward to the summer of 1995: I am now working at The Natural History Museum (NHM), which has agreed to fund a joint ten-day excavation at Pavāri with the LMNH. I am flying into Riga with my wife, Janet, who has been persuaded against her better judgement that this might be a nice way to spend part of our summer holiday. We are met at the airport by Ervīns, who picks us up in his car. It is a little blue contraption, like a bottom-of-the-range 1950s Fiat, with a two-stroke engine that sounds like a lawn mower at full throttle. The streets of Riga are still potholed, but the city is starting to shake off its Soviet greyness. Some of the splendid Art Nouveau mansions around the edges of the Old Town have been renovated, their fantastical roof lines and lushly ornamented facades picked out in clean bright paintwork. New shops and bars seem to be springing up everywhere, and the narrow alleyways of the Old Town itself are awash with tourists. Riga is starting to look like a city in the West. And yet, there is something sharp-edged and raw about the place – smart restaurants alongside dilapidated tenements, limousines with tinted windows cruising past tired and shabby people waiting for the trolleybus home; everywhere you sense that the new-found wealth is concentrated in few hands.

Eventually we arrive at Ervīns' house, on a typical mid-1980s suburban estate. He lives with his family in a small flat on the eighth floor of a crumbling, sagging tower block of such stunningly poor build quality that, at 12 years old, it already looks ready for demolition. Stepping out onto the balcony requires considerable courage. But the flat itself is a haven of civilised domesticity, and we are made very welcome by Ervīns' wife, two children and elderly mother-in-law. The evening passes pleasantly. Granny and the kids settle down to watch television – melodramatic Latin American soap operas dubbed into Latvian – while Janet and I catch up on events with Ervīns. He feels hopeful about his country, but times are hard for state employees. A curator's salary does not go very far. Fortunately, the Government has returned to him a farm in the south of the country, which used to belong to his grandfather but was seized by the Soviet regime. The farmhouse is in ruins, but the soil is good and the family can grow their own vegetables and flowers.

We spend the next couple of days at the LMNH, assembling equipment for the excavation and meeting the other members of the dig crew. They are a motley crowd. There's Anita, Head of the Geology Department (LMNH), a craggy and broad-shouldered woman in her early fifties; Valdis the taxidermist, not much short of seven foot tall, a soft-spoken giant who seems to have wandered out of some ancient folk tale; Nikolai, the entomologist, a slight and mercurial man with an encyclopaedic knowledge of moths; and half a dozen others. Nikolai has learnt a lot of English since my last visit, but unless Ervīns is around to interpret, communication with the others is limited mostly to smiles and gestures.

At last, everything is ready. In a convoy of cars, stuffed to the gunwales with pots and pans, tents, sleeping bags, spades, plaster and wrapping material, we head off into the countryside. Pavāri lies in Kurzeme, the westernmost province of Latvia. It takes us some three hours to get there, with a brief lunch stop at a dreadful, mildewed roadside restaurant that Ervīns seems to remember was quite good last time he passed through but which has gone badly downhill since. A light but persistent rain has started to fall when, driving along a featureless stretch of road with forest on either side, we spot a purposely broken sapling by the verge. It is a sign left by Valdis, who has gone ahead, marking the almost invisible entrance to the forest track we should take. A sharp turn to the right, and we are heading steeply downhill through the trees, rain-sodden branches of rowan and sallow slapping against the windscreen. Valdis himself is waiting at a bend in the track, grinning broadly. He is in his element. Then the view opens out as we emerge into the grass of a small meadow, about the size of two tennis courts, where the track ends. Behind, and on either side, the forest stands silent and dripping in the rain. Ahead lies the dark, smooth stream of the River Ciecere. We have arrived.

Valdis, bless him, has somehow managed to light a campfire despite the rain. There's hot water in the big galvanised bucket that serves as a kettle, and instant coffee and mugs are soon set out on a rickety folding table by the fireside. By the time we have pitched our tents – sturdy no-nonsense affairs without flysheets – and unpacked the sleeping bags, it is time for dinner – a stew of tinned meat and potatoes, cooked in a cauldron on the fire. Gradually it starts to get dark. As ever with fieldwork, the toilet arrangements are the bleakest part of the experience. On this occasion they consist of a spade, a loo roll and the surrounding forest. Wellingtons, a torch and mosquito repellent are essential accessories. And so to bed.

The next morning we set off on foot for the dig site, which lies about a kilometre upstream. Like almost all Latvian fossil localities it is a low, crumbling riverside crag. The explanation for this, and for the gentle rolling landscape of the whole south Baltic coast, lies in the geological history of the region.

Figure 4.2 The camp at Ciecere.

The Devonian sedimentary rocks lie draped over the southern part of the Fennoscandic Shield, a very ancient continent block that stretches from Norway in the west to the Urals in the east. At the end of the Devonian period, this Shield straddled the equator. During the 365 million years since then, the continents have danced a complex minuet across the face of the globe, sometimes colliding to raise mountain chains, sometimes tearing apart to form new oceans. But the Fennoscandic Shield has always remained an unbroken unit, and only its very edges have been affected by mountain building and folding. As a result, the sediments on its southern part have remained remarkably undisturbed. Their strata are almost horizontal, stretching for miles without folding or faulting. Furthermore, whereas Devonian sediments in other parts of the world are generally lithified (that is turned to solid stone by minerals from the ground water that have cemented the sediment grains together), this is not the case in Latvia. The outcrop at Pavāri does not consist of sandstone, but *sand*; only compaction and moisture lend it the semblance of rock. Soft sediments like these tend to weather into deep, fertile soils. Only when a river cuts into the land surface does the Devonian sediment come to light. Many of the geological strata of Latvia bear the strange, musical names of the rivers along which they were first recognised: Abava, Gauja, Amata, Daugava, Imula and Amula.

As expected, we catch Nature in the process of reclaiming our dig site. In the four years since our last visit a lot of sand has washed down onto the excavation floor and tall weeds have sprung up on the spoil heaps. Still, things don't look too bad on the whole. Even now we can see fragments of brown bone weathering out of the sand here and there. We spend the first day clearing the site, and soon discover that the vertical excavation face presents a problem. The ground slopes up steeply behind the site, and cutting into the face in pursuit of bone soon produces a dangerous overhang of soil and bushes above your head. In 1991 we had gone as far in as we dared, and then stopped. This time we will need to remove some of the overhang before even thinking about further excavation.

A particularly nasty tangle of roots, more than a metre across with a small spruce tree growing out of it, hangs threateningly above the middle of the cutting. It will have to go, but how? You would need to stand more or less on top of the root ball in order to work it loose, but nobody wants to risk plummeting into the excavation when it goes. Ervīns and two of the other men start poking at it in a half-hearted fashion, keeping their feet firmly on safe ground, but they are not having much effect. At that point Nemesis arrives in the form of Anita, who has come over from the camp to see how we're getting on. She is less than impressed, and strides up to take charge of the project. Five minutes later, after an awe-inspiring onslaught with spade and axe, only a few slender fibres are holding the root ball in place, and Anita is jumping up and down on it to tear it loose – heedless of the drop of more than 2 metres below. The lads just manage to grab her when it finally crashes into the cutting. All in a day's work, says Anita, and starts clearing away the debris.

Excavating at Pavāri is an extraordinary experience. Hammers and chisels would be the order of the day at most early tetrapod localities, but not here: the most useful tool at Pavāri is an old breadknife. The Devonian sand, pale greenish grey with occasional lumps and bands of bright-green clay, is soft enough to gouge with your fingernails. By sweeping the breadknife across its surface you can slice off a millimetre at a time, gradually working down through the sediment until you hit bone. That's when things get tricky.

As often as not, the first sign of a fossil comes when the sound of the knife suddenly changes, from the silky swish of sand over steel to a low juddering rasp.

Figure 4.3 Ervīns and Anita removing the
overhang.

When this happens you stop the blade immediately. Now a lot depends on luck. If you are lucky, you have hit the broadside of a bone – a dark-brown shadow disappearing under the semi-translucent sand – without causing any damage. If you are unlucky, you will have caught the edge of the bone and broken off a piece. Pavāri fossils unfortunately have all the robustness of soggy digestive biscuits. It is heartrending to break something that has lain safe and intact for more than three million centuries, and you try your best to be careful, but sometimes it happens all the same. All you can do is swear under your breath and carry on.

Careful scraping with the point of the knife, and brushing with a paintbrush, gradually uncover the rest of the bone. Once it has dried a little, a solution of plastic consolidant is applied. This soaks into the bone and surrounding sand and hardens them. Most of the fossils at Pavāri are 'dermal bones' – broad, thin plates of armour that lay under the skin surface in life. The bones of the human face are of this kind, but in us they have become deeply buried under flesh and muscle. In Devonian vertebrates, by contrast, the dermal bones were covered only by a thin layer of skin, and their outer surfaces are sculptured with a pattern of ridges and hollows called 'dermal ornament'. Nobody knows what purpose this ornament served, but it is very useful to palaeontologists. All the groups of Devonian vertebrates have slightly different ornament patterns, so it is possible to identify a fossil even from a small piece of armour.

Almost all the bones emerge in a single layer that undulates up and down through the sand. There are no complete skeletons, but sometimes a few elements remain articulated together. It looks as though the carcasses had disintegrated before they were washed down and buried here, maybe during a flash flood after heavy rains upstream. Many different species lie mingled in this ossuary. By far the commonest remains are those of *Bothriolepis*. Alive, it would probably have appeared to our eyes the most outlandish member of the whole fauna. Picture a smallish fish, up to about 45 centimetres in length; the hind part of the body looks fairly ordinary, though the tail fin is asymmetrical and shark-like, but the front part of the body is wholly encased in bony armour. The head is blunt and rounded, with eyes set close together on top and a small mud-grubbing mouth slung underneath. Behind the head, on either side, protrude the pectoral fins; but they have become entirely encased in bone and mimic to perfection a pair of crab's legs. *Bothriolepis* resembles nothing that lives today and has no living relatives.

Bothriolepis bones are easy to recognise from their regular dermal ornament, which looks almost machine tooled. There are hundreds of them at the site. Sometimes an articulated headshield turns up on its own, looking like a tiny replica in

bone of an old copper diving helmet. Scattered among *Bothriolepis* fossils are remains of other, rarer animals. Round scales covered in wavy, parallel ridges belong to *Holoptychius*, a large predatory fish distantly related to modern lungfishes. The scales of *Ventalepis* are large, smooth and diamond-shaped. They are common enough, but no other part of the animal has ever been found and we have no idea what it looked like. It is a salutary reminder of how little we really know about the life of the Devonian period. At times, when the bones are emerging crisp and perfect from the sand, it can feel as if the unimaginable abyss of time separating us from this vanished world is actually being bridged. But in reality all we will ever have are a few partial glimpses of its weird denizens, together with the knowledge that many more throng the shadows just beyond the reach of our vision.

We begin to find bones of the tetrapod, *Ventastega*. They are very distinctive, with an ornament of irregular meshing ridges radiating out from a central point, and there are quite a few of them. So far so good. But the bones do not represent all parts of the skeleton equally. Something, probably the water current at the time of burial, has sorted the bones. Most of what we find are lower jaws and broad fan-shaped collar bones; in one place three tetrapod jaws and one fish jaw lie stacked together like spoons in a drawer. This is a problem, because we want to know more about the anatomy of *Ventastega* and it won't do just to keep finding the same bits over and over again. I score one success quite early on. Patient scraping at a small patch of bone in the sand gradually reveals a complete lower jaw, showing for the first time the delicate rear end complete with the joint surfaces of the jaw hinge.

As the days go by, the efforts of each excavator generate a little hollow in which he or she sits crouched, scraping away at the dig face. The work is not heavy but is nevertheless backbreaking. The cool, damp sand chills your muscles and makes your limbs and back stiffen in protest by mid-morning. Every so often you need to stop and stretch your legs. There is plenty to look at. The Soviet period saw horrendous environmental damage on a local scale, but the wider countryside was spared the intensive agrochemical farming that developed in the West. As a result, it is swarming with life. Every puddle seems to have a frog in it, and clouds of butterflies haunt the forest edge along the Ciecere. It is shocking to realise how threadbare our own countryside has become by comparison. Nikolai the entomologist seems to be on first-name terms with all the local insects and will cheerfully identify anything that passes, adding that the White Admiral, or whatever it is, 'is actually rather common.' Now and again a kingfisher darts by like a whirring streak of turquoise against the dark river. There are larger animals too. We often find the wallows and tracks of wild boar in the forest, but the pigs themselves stay out of our way. One day a black stork flies over our

Figure 4.4 (*Opposite*) Hard at work.

meadow. Not that all this biodiversity is an unmingled joy. Our camp is invaded by wasps and several people get stung. Even a hornet turns up one evening, large as a thumb, but it flies off into the forest again to everyone's relief. Nikolai gets bitten by a tick and spends the rest of the week worrying about catching meningitis.

Large areas of bone bed have now been exposed. Sand is starting to get everywhere, even the lunchtime sandwiches of dark rye bread are beginning to taste gritty, but still we are only finding the same bits of *Ventastega*. Jaws and collar bones, collar bones and jaws; will we really discover nothing new? Then our youngest digger, the 14-year-old son of one of the museum geologists, begins to uncover a large flat bone. It is kite-shaped, the size of a man's hand, and there is no doubt what it is: an interclavicle of *Ventastega*. In life, it would have formed part of the shoulder girdle, lying in the middle of the chest between the two collar bones. Whoops of delight erupt from the crew. This is a major new discovery. What's more, the shape of this bone varies among the different Devonian tetrapods. We can already see that our new find resembles the interclavicle of *Acanthostega*, but differs greatly from that of *Ichthyostega*. It may give a clue to the relationships of these animals.

At last the 10 days are nearly up, and we begin to prepare the bone layer for removal. This is a sensitive stage. Even though consolidated, the bones are still fragile and would simply shatter if you tried to lift them without support. Carefully we cut narrow trenches, about 2.5 centimetres wide, into the bone bed, dividing it into manageable blocks and trying to damage as few bones as possible in the process. While the blocks are still in position, we wrap their sides round with strips of sacking dipped in plaster of Paris, moulding these bandages as closely as possible to the sand. Then, once the plaster has dried, the block is gently prised from the underlying sand and lowered onto another piece of plaster-soaked sacking, which will form the base of the package. Everybody holds their breath; if the sides of the block have not been carefully wrapped, the whole mass of sand and bones can suddenly collapse and fall out when it is lifted, leaving the excavator with an empty plaster bandage in his hands and tears in his eyes.

The easiest way of getting the blocks from the dig site to the camp is to carry them by boat along the Ciecere. Ervīns has brought his trusty old rubber dinghy for the purpose. Unfortunately it is not as airtight as it once was, and there is some speculation as to whether he will be able to complete the 1-kilometre trip without deflating and sinking with his precious cargo. At least this time he only has to contend with rocks and fast-flowing water en route; in 1991 he got caught in a mist net that some ornithologists had stretched across the river to catch and ring kingfishers. He makes it to the camp afloat, but only just. We have to pump the boat up again for the

Figure 4.5 Blocks of bone bed in plaster bandages.

return trip. Meanwhile the rest of us begin packing up our things, shaking the sand out of our sleeping bags and trying to find that one clean shirt lurking at the bottom of the rucksack. The mood is both happy and slightly melancholic. We have made some good finds and we will be sad to leave, but everybody is keen to get back to beds and flush toilets. On the last evening I bring out a bottle of malt whisky, which meets with general approval. As a return gesture the Latvians decide to spike some tinned fruit juice with the remnants of the technical alcohol we use for dissolving consolidant. Janet and I decline to partake, fear of meths poisoning getting the better of our social graces. The next morning we are pleasantly surprised to discover that nobody has died or gone blind during the night.

And so we pack our gear into our cars and drive home, leaving the meadow and the river and the kingfishers to take care of themselves. Once back in Riga, at the LMNH, the next phase of the project will begin. Slowly and painstakingly, working with needles under a microscope, the bones in the blocks will be freed from the embrace of the sand. It will take several years, but we can already get an idea of the results by looking at the 1991 material, which lies carefully cleaned and arranged in drawers in the Geology Department of the LMNH. There are jaws, scales, armour plates, all sere and brittle like withered leaves, but so perfectly preserved that it is possible to see how the teeth were shed and replaced or how the crab-like 'arms' of *Bothriolepis* articulated with its trunk armour.

So what have we learned from our 10 days at Pavāri? Well, for one thing, that a hole in a nondescript river bank in a small European country can prove to be a window on a lost world. We know that the Earth holds more of these treasures; who can say what unregarded corner the next great discovery will come from? As for *Ventastega*, the 1995 material is giving us a much clearer picture of its anatomy. It is beginning to look like a very fish-like and primitive animal, maybe even more so than *Ichthyostega* and *Acanthostega*. Once this information has been fully analysed, it should tell us a lot about how and by what steps fishes evolved into the first tetrapods.

We can never bring these animals back to life, or understand them fully, but their remains are not wholly mute. Through careful study they can be made to speak to us – if sometimes in riddles – about the lives of our own ancestors in an unimaginably distant past. Not bad for a bunch of old bones.

Chapter five

Brains in Abu Dhabi's Desert

Peter J. Whybrow

The Sahba is one of
the great and long-dead
rivers of Arabia, having a
total length of more than
500 miles from its head
in the flanks of 'Alam in
the central highlands of
Najd to its mouth in
the Persian Gulf.

The Empty Quarter,
H. St. John Philby, 1933

Another tropical day dawned on the river, meandering eastwards from its source in the mountains of western Arabia. Hippos had finished their nocturnal feeding in and around the papyrus reeds. They eyed the crocodiles, who, in turn, were eyeing their lunch; perhaps one of the horses drinking from a pool where a family of elephants with down-turned tusks had recently watered, or perhaps some of the numerous, sluggish catfish. A large, long-necked turtle surfaced and disturbed an egret that was stalking small, barbel-like fish in the shallows. Here, amongst the sands, freshwater mussels filtered their food. Inland, amongst the acacia-like trees, an ostrich had laid a clutch of thick-shelled eggs and a carnivorous badger sniffed at them, hoping the chicks would soon hatch.

No, not an account of an African safari, but a scene in the United Arab Emirates about 8 million years ago during the Miocene epoch.

In the Old World the Miocene time segment (from 24 to 5 million years ago) documents important evolutionary changes to land animals and, although their bones are rare, they recognisably belong to modern groups such as elephants, horses, hippos, birds and reptiles. Unbelievable though it might seem to us today, Arabia also had a rich fauna in Miocene times. We know this largely because of recent discoveries in the Emirate of Abu Dhabi by an international team of 42 scientists, 18 of them from The Natural History Museum (NHM) in London.

Every year since 1989 a field team led by myself and Andrew Hill of Yale University, with occasional help from other palaeontologists, have collected fossils from sites along part of Abu Dhabi's Arabian Gulf coast, an area previously unknown to palaeontologists. Our discoveries are important new fossil records for the whole of Arabia.

Before the mid-1970s, the main thrust of research about the relationships of Miocene animals to their living relatives took place in East Africa and south-western

Figure 5.1 Artistic reconstruction of life around
Abu Dhabi's ancient river 8 million years ago.

Asia. In Kenya on the eastern shores of Lake Victoria, the Provincial Governor in 1909
found fossilised bones that were sent to the, then, British Museum (Natural History) for
identification. They proved to be from an early relative of elephants, and the discovery
prompted several palaeontological expeditions to other remote parts of eastern Africa,
where further discoveries were made famous by members of the Leakey family. In Asia,
Miocene vertebrate fossils similar to those discovered in Africa have been collected
from the foothills of the Himalayas since 1830 and more recent work there since 1973
by palaeontologists from Yale and Harvard Universities in the USA, collaborating with
the Geological Survey of Pakistan, has resulted in an amazing collection of over 20 000
fossils.

During this early period of discovery and exploration, Arabia seemed to be
completely barren of Miocene vertebrates. This gap in our knowledge was vast in terms
of both geological time and the geography of the region during the Miocene. An area
that includes all the countries of the Arabian Peninsula, together with Jordan, Syria,
and Iraq west of the River Tigris, occupied a central position between the better-known
Miocene localities of Africa and Asia. Puzzlingly, though, it seemed to lack fossils of
animals that may have migrated 'out of Africa' millions of years ago. This was
especially frustrating as it was known that before and for most of the Miocene Arabia

was part of Africa and not joined with Asia. Geological evidence from the numerous oil wells in the region indicates that, about 25 million years ago, the Mediterranean was connected to the Indian Ocean via a seaway through what is now Mesopotamia and the Arabian Gulf. The Arabian continental plate gradually moved away from Africa to form the Red Sea basin and, in so doing, closed the seaway in the region of Qatar to form the first landbridge between Arabia and Asia about 19 million years ago during middle Miocene times.

Our lack of knowledge of Arabia's past animal life changed in 1974 when, rather like seeing the dark side of the Moon for the first time, a whole fauna of fossilised mammals was found by myself and another NHM palaeontologist, the late Roger Hamilton, in eastern Saudi Arabia. One of the animals was a primitive type of elephant whose teeth and bones had been recognised as long ago as 1930 from the Baluchistan region of Pakistan. The fossil-bearing rocks in Saudi Arabia, dated at about 19 million years, were of a similar age as those of the Pakistan elephant locality. Therefore, the elephants' ancestors might have used the first landbridge across the Arabian Gulf to arrive in southwestern Asia. This important discovery prompted a reconnaissance by myself, between 1979 and 1984, of similar aged rocks in Qatar and the United Arab Emirates.

In 1979 I travelled the long road from Saudi Arabia to Abu Dhabi through Qatar where at the United Arab Emirates (UAE) border goats sustain their life by eating the immigration forms. With a Qatari driver we crossed the vast salt flat of Sabkhat Matti, and through soft sand speeded to the top of Jebel Barakah, the highest point in Abu Dhabi's Western Region, at 63 metres (jebel is a hill). In a few hours I discovered fossils of crocodiles, horses and cattle-like mammals. Two years later the United Arab Emirates University helped my return to Barakah with a vehicle and a young Egyptian driver. Our journey did not start well. Mohammed was arrested for speeding and non-payment of previous fines. After I had bargained with the police for his release, we set off at an even greater speed through the heat haze along the shimmering coast road where the embracing horizon was never anything other than totally flat. While munching pancake sandwiches for our on-the-road breakfast, I puzzled over his questions concerning my mental health. Suddenly, I realised that the grey, glutinous sandwich fill had been cooked sheep brains. Now, some locals believe this delicacy increases mental prowess whereas others think that eating brains, especially of sheep, makes the person as unintelligent as the animal.

Onward we drove, the journey becoming longer and louder, as repetitive Egyptian music droned from Mohammed's tape recorder; the air conditioning didn't work and the heat was oven-like. 'A way to keep cool,' he inexplicably said as all

Figure 5.2 The 20-metre-high sea cliffs at Jebel Barakah. Fossil elephant remains have been discovered in the sandstones near the top of the cliff.

windows were suddenly shut. Ah, I thought, a test of a northern European. My body's biology started to melt-down as streams of sweat merged to flow from my finger tips. 'Ready?' he then queried, 'Open windows!' Soon I was freezing as sweat immediately evaporated from the hot blast of desert air.

We continued this efficient technique for several hours. But why, I wondered, was the petrol tank gauge still showing full? The vehicle soon spluttered to a halt. He had spliced a wire from his tape recorder into the vehicle's electrics. While he hitched a lift to the nearest petrol station I sat by the dead vehicle waiting to be mummified by the desert sun: isolated, dehydrated and mouthing expletives to the desert about Egyptian drivers I wondered if my mind had already been affected by eating brains of sheep.

Gradually the topography changed from totally flat to ever so slightly undulating. This change was marked by the appearance of huge advertising hoardings

Figure 5.3 The lower jaw of the Barakah
hippopotamus – as found.

beckoning fatigued traveller's westward to the Dhafra Beach Hotel. Imagining a building like the multistorey hotels in Abu Dhabi, I became worried when even at 10 kilometres from its location nothing of its structure could be seen. After all, it was said to have 300 rooms, a restaurant, squash and tennis courts. It's almost flat around there – it should have stuck out for miles around. But, no, the Dhafra Beach was seemingly invisible. At just 1 kilometre from the hotel an explanation of its mirage-like-joke emerged; camouflaged with sand-coloured paint, concealed by eucalyptus trees as high as its three storeys it totally merged with the desert. It is a modern oasis. Built in the 1970s to accommodate European and American oil engineers and managers developing the nearby refinery and oil terminal it was to be our 'field camp' for the next 10 years. Its promotional literature proclaims a 'one storey building designed to resemble a fortress for the public area [sic]' and 'The perfect spot to run away for some peace and privacy. It's so seclusive, nobody will bother to look for you here.'

I began work. Near Jebel Barakah I plodded around finding only scraps of unidentifiable bone and began to worry. Much effort, time and money (at that time the huge sum of £1000) had been allocated by NHM to this initial exploration. Returning with nothing but stories of sheep brains might have stopped all further Arabian work. Suddenly, while crawling over the yellow-brown desert surface, the bluish enamel from a row of teeth poking through the Miocene sand caused me to rub my sand-filled eyes with disbelief. With brush and trowel, excavation after 8 million years of burial revealed an almost complete lower jaw of a hippopotamus – the first from the Arabian

Figure 5.4 The Barakah hippopotamus after conservation.

Miocene. Its presence, along with the 1979 collection, proved it had been buried in sediments deposited by an hitherto unrecorded fossil river system.

Close to our field camp lies Jebel Dhanna – a dome created by huge volumes of 600-million-year-old salt breaking through younger surface rocks. Salt, gypsum, ancient granites, iron ore and sulphur – mined by the Portuguese in the seventeenth century – formed a contorted mass of black, dark-green and muddy-brown material, a geological dustbin upon which no soil has formed and therefore no vegetation exists. North of Jebel Dhanna lies a salt dome island, Sir Bani Yas, named after the powerful Bedouin tribe who controlled the interior from the nineteenth-century 'Pirate' or 'Trucial Coast' and whose descendants are now the ruling families of Abu Dhabi and Dubai. South from the coast the desert scene is, at best, bleak. The UAE army use most of it for war-games; abandoned tank shell cases become new homes for scorpions. It is neither a rocky nor a sandy desert. Gradual but consistent erosion of the soft Miocene rocks by wind and some rain over thousands of years has flattened the ancient coastal plateau to today's minimal topography. It is as if a thick wool blanket had been languidly draped over the area: no natural landmarks; no naturally growing trees; few shrubs; some salt-tolerant plants; some lizards; some migrating birds and sometimes the rare desert hare.

No one but a geologist would voluntarily visit this place. In Spring, it can be mind-blastingly hot: walking becomes a plod; breathing is laboured and tiny flies seek moisture from nose, eyes and lips – there is no wind and no shade. On other Spring days, however, when the air lacks any humidity, and in the early morning when the sun is low, the desert appears to pulse with exuberant artistry. A sky of an astonishingly brilliant shade of blue overlies a velvet-like silence, punctuated only by the song of desert larks, and the chattering wings of huge dragonflies. The light has an icy clarity, and the previously dull desert is briefly coloured with merging shades of red, brown and yellow speckled here and there with the iridescent green of plants.

How do we find fossils? First, experience helps. The cliché that the best palaeontologist is, amongst other things, the one who has seen the most fossils is true. Second, a knowledge of zoology helps a great deal because we are looking for shapes that are not minerals, not sedimentary structures nor weathered pebbles, but skeletal remains. A vertebra or a rib just poking out of the rock has, to us, a recognisable form. Sometimes we are totally wrong. Field identifications, especially of bone fragments, can be suspect because comparative material is unavailable, and the bone is either attributed to the wrong animal or to the wrong part of the body. Finally, there is the search itself.

Figure 5.5 Neither a rocky nor a sandy desert –
the interior of Abu Dhabi's Western Region.

A colleague once said that a search for fossils appears to consist of a lot of aimless wandering. We note the types of rock exposed and try to imagine what the environment was like millions of years ago. In fact, clues are being looked for: this clay layer is no good for large fossils as it was deposited in very calm waters; this coarse-grained, bedded sandstone is no good as it seems to be river bed deposit. But here is a

Figure 5.6 Hunting for fossils on the island
of Shuwaihat.

pocket of fine-grained sandstone that may have been a bar in a meander of the river. So, below where this sediment is exposed, we start to crawl up the slope with our eyes about 10 centimetres from the ground – not much fun in a high wind. Soon, recognisable shapes appear – tiny fish teeth, pieces of crocodile bone and complete rodent teeth all winnowed out of a fossil-bearing layer by recent erosion. After finding this layer, we must find how far it extends and, invariably because the sediments are river-channel deposits, it peters out into seemingly barren sandstones.

Besides collecting and identifying fossils, other aims of our research were to collect rock samples for analysis, and measure the thickness of the rocks exposed on the jebels so as to correlate the fossil-bearing layers over a large area of the region, about 16 000 square kilometres. This work involved identifying certain types of sediment, such as sandstones or clays, on 50-metre-high jebels down to the nearest centimetre. This is extremely difficult when the cold north wind – the Shamal – from the high plateaus of Iran and Russia is driving sand into exposed flesh while I slip down the cliff into the sea. The fleshless fossils lying on the surface are also being abraded by sand-laden wind. Although some may be lost to science, others were being revealed by the wind even as we searched.

In 1989, the President of the UAE, Sheik Zayed, came to the Dhafra Beach

and the hotel staff shunted the residents (us) from the wing he was to occupy into another. Security was interesting. The army had dug foxholes around the hotel's perimeter and they assiduously loomed out of the dark to confront weary palaeontologists returning from fieldwork. Whatever the Arabic for 'Sod-off' is, we were told to do so on trying to enter the hotel. 'Come back later' turned into three hours of waiting and, being January, the night temperature was freezing – no wanderings around the warm, starlit desert to temper the delay.

Andrew, Sally McBrearty (an anthropologist from the University of Connecticut), Walid Yasin (Al Ain Museum's senior archaeologist) and myself escaped from fortress Dhafra Beach and headed for Ras Dubay'ah. 'Ras' means headland and maps indicated that it was a short drive of a few kilometres from the main road. In fact, Ras Dubay'ah was surrounded by potentially dangerous salt flats – sabkha. This stuff looks innocuous. Its flatness suggests that a vehicle can be easily driven over it at great speed. But walking on it is like walking over dirty brown cornflakes except when the thin, crusty, apparently hard surface breaks and the underlying crystalline mud starts to suck and pull the walker into porridge-like black muck. Vehicles have disappeared into this stuff.

We skirted the worst of the sabkha but the dry ground ran out some 5 kilometres from our goal where it was interspersed with water-filled channels left by the receding high tide. Andrew elected to walk a few hundred metres, stomping along and pretending his weight equalled that of a 2-ton Toyota. He soon gave up. Walid was in the lead vehicle. For him, driving over sabkha was more exciting than finding archaeological sites. He zoomed off and I followed at some distance. We had neither shovels nor tow rope and my vehicle had almost bald sand tyres. It was thrilling. Instant decisions had to be made. Must keep out of Walid's tracks; whoops – traction lost, shall I stay in second gear or go for low four-wheel drive? How deep is that channel, skirt it or just blast through it? Damn, wheel spin in a hole, grind the gears and pray. Panic, bash through a flooded channel. I'm blinded, windscreen covered in goop. Why are the indicators flashing? Of course, left-hand drive, wipers on right. Oh ****, the foot of Dubay'ah is covered with thick, soft, dust-like covering. I spin the vehicle around, search for hard ground and park down hill.

We had arrived and felt like real explorers safely achieving an impossible objective. Look there! We, the first palaeontologists to visit Dubay'ah, had seen fragments of bone eroding from its sandy red cliffs. We scattered, each wanting to be the first to find more evidence of Miocene mammals, especially hominoids, that had once dispersed from Africa or Asia through Arabia. Sally discovered a fragmented elephant skull, Andrew found a rhinoceros tooth and, perhaps, a vertebra from a big

Figure 5.7 (*Opposite*) Peter Whybrow (left) and Andrew Hill crawling over Miocene rocks seeking fossils.

sabre-tooth cat and he and I found pieces of bone identifiable as giraffe, primitive horse, crocodile and turtle. Walid shouted to us. He had found a badly preserved elephant tusk that he wanted collected. He seemed very keen on this and was disappointed when told that it was unidentifiable and would, in any case, fall to bits. Downcast, he disappeared over the hills of Dubay'ah. There was another, more distant shout from Walid, who could just be seen peering intently into the ground. With brush and dental pick he was removing sand from around a fractured but complete molar tooth of an elephant. How he saw it was a puzzle. A detailed search over the flat, football pitch-size hill, indicated that Walid had found the only fossil. We packed it with sand in Walid's toolbox and, mindful of our return journey over the sabkha, left further work at Dubay'ah to another day.

After our usual slaking of thirst in the Falcon Bar of our field camp, we joined Walid for dinner. An off-the-cuff remark along the lines of 'Pity the President does not know about us finding a fossil elephant' got Walid excited. He chatted, in Arabic, to various security people. He shrugged unknowingly when we enquired 'What's happening?' and it was when trying to decide on which of the eight puddings was not made with 90 per cent sugar that we were unexpectedly summoned. A regal security man poked his mobile phone aerial into the box of sand containing the elephant tooth (did it contain a 'device'?). Andrew and I were given permission to hump the heavy box to the third floor of the wing from which we had been ejected a few days before. The door to a smallish room was flanked by immensely tall Somali bodyguards adorned with ammunition bandoliers. Our treasured fossil was carried to where the President of the United Arab Emirates, His Highness Sheikh Zayed bin Sultan Al Nahyan, Ruler of the Emirate of Abu Dhabi was sitting. Twenty or so members of the Presidential group were similarly seated, ceremoniously tapping the floor with long camel sticks over which Andrew and I tripped with humble apologies. A sinuous trail of desert sand from our box added more colour to the regal carpet as we placed it with our treasured elephant tooth on a luxurious walnut table in front of His Highness.

Seated, we peered at his entourage from around a mountainous platter of fruit while Walid, acting as our translator, described our work, detailing the number and variety of fossilised animals we had found; how a river had once flowed just a few kilometres from the hotel and how the climate must have changed since geological times. In turn, His Highness told of legends that also described such a river that, to us, must have existed more recently; this in itself was important information. Departing our audience, Andrew remarked 'We have just visited the President of one of the world's richest oil-states. You haven't shaved for two days, your trousers are covered with sabkha mud and I wasn't wearing any socks.' Such is the stuff of field work.

In January 1991, I was back in Abu Dhabi. At about 3 a.m. on the morning of 17 January, I was awakened by the drone of high-flying aircraft, presumably from forward bases in Oman. The bombing of Iraq had begun. I 'kept my head down,' the advice given over the BBC World Service. That night, bits of Scud missile landed in Riyadh less than 800 kilometres west of Abu Dhabi. Saddam Hussain had threatened to obliterate the oil-loading terminal at Jebel Dhanna, the place I was to revisit.

With field equipment and, importantly, a military pass, I passed through numerous checkpoints and arrived at Jebel Dhanna's very busy Dhafra Beach Hotel. Now the hotel's prime concern was assisting with the supply of food to the Allied Coalition. Resident were older, short-haired Americans of a senior rank. A phone call from my friend Nasser Al Shamshi, in charge of Abu Dhabi Company for Onshore Oil Operations (ADCO) Government Relations, told me he would visit the next day. 'I have made introductions for you,' he said, clutching his mobile phone as we sped towards army headquarters in his new Mercedes. A vast carpeted room with about ten army top-brass and a few civilians (that is nationals in Arab attire) greeted us. Seated in deep chintz chairs we drank numerous cups of Arabic tea and coffee. At last, a question to me in English, 'Was Thesiger a spy?' asked the senior man about this enigmatic British traveller and the second person to cross the Rub al' Khali desert. Thinking quickly I said 'He was invited by the SAS to Yemen in the 1960s during the Egyptian-led insurgency into Royalist Yemen.' This reply seemed to go down well. A hub-bub of Arabic followed as Nasser told me that I had confirmed that Thesiger was a spy and that the UAE was right in deporting all its Yemeni workforce.

Wondering how the Foreign Office might analyse this perception, we left, but along the road an army vehicle sped past and halted us. Oh ****, I thought, perhaps my answer proved that I also was a spy. There was much Arabic talk and I was introduced to one of Nasser's many cousins. 'Come to lunch,' he said, and even more coffee was supplied in yet another vast carpeted room. A large plastic table cloth became a floor cloth and other locals poured through the door to join us squatting around a delicious feast of traditional seasoned mutton with rice. We were about to leave muttering 'Al-Hamdu Lilah – I have had sufficient, God be Praised', when a military man stated that I had not yet had the best part of the carcass. He split the lower jaw in two, turned over the skull to expose the roof of its mouth, and using the jaw bones to lever the skull apart, exposed – nothing. A stunned silence. Much jovial argument then took place on how this sheep had managed its life without a brain – it was thought to be an Australian import – and I was relieved that I did not have to have more of this delicacy. Locally, this event has become a puzzling legend, 'Do you remember the day of the sheep with no brain?'

Nasser left for Abu Dhabi and I, alone in the war zone, went to Shuwaihat, an island linked to the mainland by a rocky causeway built by British Petroleum in the 1950s to transport a drilling rig. Andrew and Walid had briefly visited the island and declared it a palaeontological paradise. The island itself was idyllic. The contrast between its sienna and umber sea cliffs with the clear and truly aquamarine waters of the Gulf was amazing. Such a place should have been preserved for all. Unlike Europe where everything is green and where the light is diffuse, the quality and clarity of Arabian air and light exhibited in detail the geological structure of Shuwaihat. Also visible was a new military communications aerial. Avoiding this, I checked out the area where Andrew had found fragments of an elephant's lower jaw the previous year. Suddenly from around the seaward side of a jebel at close range and at a height no more than 20 metres clattered an US Marine attack helicopter with its gun – Puff-the-Magic-Dragon – aimed directly at me. What to do as thoughts of friendly fire focused my mind – wave my hat, lie on the ground or run? The thumbs-up sign worked and as I walked around the jebel I saw the reason for the security. Along the horizon was an American armada about to practise an amphibious landing in the area where Sabkha Matti meets the Arabian Gulf. Their very visible presence – every truck driver using the Abu Dhabi Arabia route must have seen the warships – apparently led the Iraqis to believe an assault on Kuwait from the sea was being planned.

This field trip was not unproductive. Highlighted by the setting sun, fossilised mammal teeth had been exposed by a few days of high winds. I excavated the jaw of a primitive horse and a colleague from the Muséum National d'Histoire Naturelle in Paris has since described the fossil as a new species, called *Hipparion abudhabiensis*.

In contrast with other localities, our collection of fossils from Shuwaihat is large and suggests our frequent visits were because it was such a wonderful place – are the number of specimens biased? No, Shuwaihat has since proved to be the most productive place in Abu Dhabi for Miocene vertebrate fossils and its significance to both Arabia and to studies of the dispersal of Old World mammalian faunas was, in part, due to the elephant remains discovered there by Andrew. In January 1992, Gill Comerford, from the NHM's Palaeontology Conservation Unit, and Andrew began the removal of the soft sand from around the elephant's lower jaw. This revealed part of its skull with fragile tusks and teeth in place as well as a leg bone, its shoulder blade and a few ribs.

During excavation as the freezing Shamal chilled our own bones it became obvious that the bones of the elephant were concealed over a wider area

Figure 5.8 Miranda Bernor and Peter Whybrow starting to excavate the Shuwaihat elephant skull.

beneath at least half a metre of soft Miocene sands. In Spring 1992, a team of seven continued excavation of the skull and more ribs, vertebrae (parts of the backbone), leg bones and other fossils – crocodile, pig, fish and turtle remains – were found in the deepening excavation.

We have published two new names for the Abu Dhabi Miocene rocks. The Baynunah Formation, after the local name for the region where these rocks are best exposed, and the underlying Shuwaihat Formation, revealed by marine erosion at the base of most of the sea cliffs. The Shuwaihat was deposited about 13 million years ago when Abu Dhabi had a desert climate not unlike the present day – dunes, sabkha and very little rainfall. A river then eroded most of the Shuwaihat Formation and deposited its sediments in numerous channels separated by low sandbanks. The river itself may have been an ancestral extension of the present day Tigris/Euphrates delta or, more

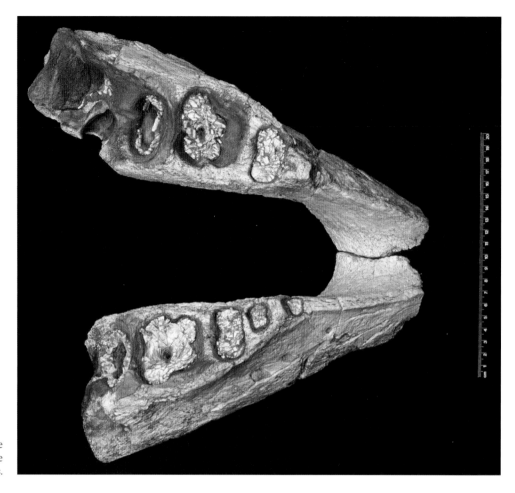

Figure 5.9 Only the third example known of the jaw bones of a primitive elephant; its teeth are missing. Scale in centimetres.

likely, the eastern part of a Saudi Arabian system that flowed north-eastwards along a trough – a line of geological weakness – that now forms part of Sabkhat Matti.

The climate changed dramatically between Shuwaihat and Baynunah times. The earlier extremely dry conditions gave way to a river system. This new environment provided a habitat for freshwater molluscs, snails, fish, aquatic reptiles, birds and mammals. Evidence of a permanent flow of water in this river is provided by remains of large freshwater turtles and crocodiles, but the presence of catfish suggests that flow was sometimes sluggish or intermittent in some channels. Occasional fast flow is indicated by coarse sands in channels and by the disarticulated and fragmented state of some of the fossil bones.

Temperatures were warm during Baynunah times, and hard carbonate layers preserved in the sediments provide geochemical evidence that the climate was semiarid, with an annual rainfall of no more than 75 millimetres. The vegetation, so

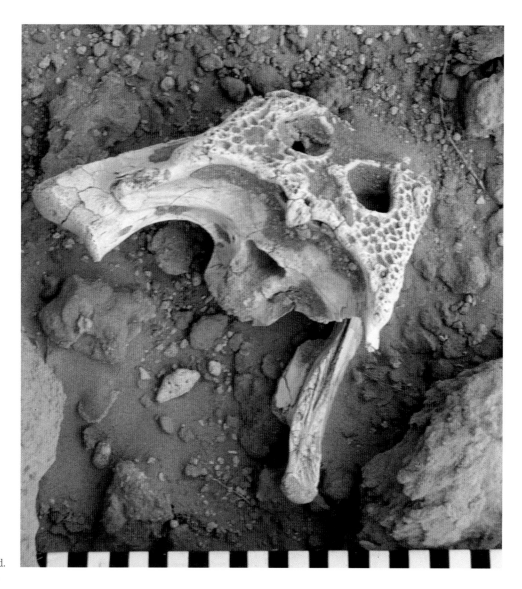

Figure 5.10 Part of a crocodile skull – as found.

the fossils tell us, consisted of a mixture of grass, shrubs, and trees, including *Acacia*. Trees and shrubs were probably concentrated near the river banks, while a more open grassy vegetation grew farther away from the river itself. This habitat supported a rich and diverse group of animals, including ancient forms of elephant, hippopotamus, horse, antelope, wolverine, hyaena, and sabre-tooth cat.

The Baynunah fauna now comprises over 900 fossils ranging from complete bones to identifiable bone fragments. They represent 43 species of vertebrate belonging to at least 26 families. Our team has named two new species and one new genus: a catfish, a gerbil named *Abudhabia baynunensis* and the primitive three-toed horse. There are three species of fish, three species of turtle including both terrestrial and aquatic forms, three species of crocodiles, including a gavial, and two species of birds. The Baynunah fauna indicates a markedly different environment and climate in Baynunah times than at present, or during preceding Shuwaihat times. The fossil-bearing sites in Abu Dhabi are unique to Arabia – in no other area of the peninsula can such fossils be found. This fact is especially pleasing to the sponsors of our work. Initial support came from the Al Ain Museum and British Petroleum, and since 1991 the ADCO has been the major sponsor.

I have frequently used the past tense in describing places such as Shuwaihat and Barakah. This is not simply to describe past events. The pace of progress in the UAE is remarkable. An area of untouched desert one year can, the following year, be the site of a vegetable canning factory or low-cost housing for fisherman. The fossiliferous areas we easily visited in 1991 by turning into the desert anywhere off the main road are now inaccessible due to newly erected wire fencing, hundreds of kilometres in length. Within these huge compounds ploughs break the ancient desert surface so that irrigation pipes can drip water to hundreds of newly planted shrubs and trees. The tops of some jebels have been levelled to build water towers to supply the irrigation systems or to build defensive missile sites. Around others massive bulldozers have dug through the sabkha to construct circling berms to dam, it is said, water from infrequent rains. It is now almost impossible to visit many of our fossil sites and it is rumoured that idyllic Shuwaihat Island might become a safari park.

There is always the hazard that description of an 'unspoilt and unknown' region, especially if it lies near the Tropics and has beautiful unspoilt shell-sand beaches, will speed its demise into the popular pit of 'been there, seen that, done it.' But the Western Region of Abu Dhabi will, for me, always be a special place, for its desert and for the opportunity to look into Arabia's past and discover the remains of animals that once lived around its ancient rivers.

Chapter six

Thomas Hardy, driver ants and some West African fossils

Steve Culver

Undyed merino socks are recommended. Light woollen pyjamas should be worn at night and a flannel dressing-gown is occasionally very useful.

Handbook on the West African Colonies,
Overseas Settlement Office, 1920

It was my fourth day in Sierra Leone. Bullets ricocheted from the side of the Faculty Guest House. I lay flat on the floor of a first-floor bedroom, hands over my head, clad only in briefs as the mosquitoes whined over me, taking turns to pierce and suck the blood from my body. The thought quite naturally crossed my mind, 'What have I done to deserve this? Why me? I'm a palaeontologist. I just want to teach palaeontology. I'm only 25. I'm too young to die.'

How had I got myself into this predicament? Just six months before, in October 1976, the day after I had completed my PhD in micropalaeontology (the study of small fossils) at University College Swansea, I had signed on the dole. The person behind the counter seemed to think that my geological background meant that I should be well qualified for a job at Port Talbot steelworks, although I couldn't quite see the link. I was qualified as a palaeontologist and had wanted to work as one since I was seven years old. (When I was nine I was runner-up in an 'I want to be competition' with my essay on palaeontology; the winner wanted to be a gardener.) I had been told by numerous people that I could never be a palaeontologist because competition was too severe and there were too few jobs around. The careers advisor at Swansea suggested I become a British Rail stationmaster. 'It's not all blowing whistles, you know' was his major selling point.

These apparently discouraging comments had quite the opposite effect and I quietly determined to prove the nay-sayers wrong. So imagine my joy when I was offered the job of Lecturer in Geology (with emphasis on palaeontology and associated 'soft-rock' subjects) at Fourah Bay College, University of Sierra Leone.

After a flurry of lecture writing, and visits to the dentist and doctor for extensive cautionary check-ups, I drove slowly through the freezing December snow along the M6 motorway to deposit my stereo, books and heavy winter clothes at my parents' house in Blackpool. I was due to fly to Freetown, Sierra Leone's capital city, in early January for the start of the new term. But just before Christmas I was informed that civil unrest had broken out across Sierra Leone, precipitated by an anti-President student rally at the university college I was about to join. The university was closed down. I was forced to remain on the dole in Swansea until mid-March 1977, when I was happily informed by the Ministry of Overseas Development that it was safe to take up my lectureship in Freetown and begin my career.

For a naive young man who had never inhabited the real world outside of academia, arriving in West Africa for the first time was an awesome and overwhelming experience – as every travel writer says, 'the smells, the sights, the sounds!' The Boeing 707 touched down at Lungi Airport some 30 kilometres north of Freetown. On coming to a halt, a local airport official dressed in flip-flops, cut-off jeans and a T-shirt proclaiming 'No woman, no cry' walked down the centre aisle extravagantly spraying an aerosol fly-killer over all of the passengers. Duly sanitised, I exited the air-conditioned plane and immediately melted in the stiflingly hot air laced with the delicate aroma of aviation fuel.

I had been warned that customs and passport control could be an interesting experience and my apprehension was not eased when I was surrounded by several Sierra Leonean youths all jostling to escort me through the formalities. I chose one carrying a sign emblazoned with the name of a major car-hire company. He pushed his way through the throng with extensive use of the sign as an offensive weapon. Within five minutes I had collected my single suitcase and, without stopping once, I was ushered out onto the airport forecourt. It was only when I gave the sign-bearer 5 leones (£2.50) for his troubles that I realised I had unknowingly bribed my way through customs and passport control. I had heard that small payments ('dashes' in local terminology) could ease one's way but I had not planned to use that approach immediately on my arrival in West Africa! It was a good 40 minutes before my fellow passengers, legally entered into Sierra Leone, joined me on the bus to Freetown.

The bus trip, including a 45-minute ferry ride across the Rokel River estuary, the finest deep-water harbour in West Africa, terminated in the centre of Freetown at the Paramount Hotel, just down the road from the City Hotel, immortalised by Graham Greene's *The Heart of the Matter*. Thankfully, two of my new university colleagues met the bus and drove me to the Faculty Guest House situated on the margins of the Fourah Bay College campus some 460 metres above sea-level. The college buildings nestled on the slopes of the Freetown igneous complex, intruded deep into the crust some 200 million years ago when Africa began to split apart from the Americas. The intrusion, which forms a magnificent 30-kilometre-long, tree-covered mountain range rising steeply out of the sea to a height of over 900 metres, gives Sierra Leone its name 'Lion Mountain'.

The Guest House, a three-storey structure built during the 1960s, was to be my home for a week or so until my own university accommodation became available. The landlady greeted me warmly and assigned me a room on the ground floor. Unfortunately, the building had been ransacked by various rioting mobs over the past few weeks and, although the door to my room would close and lock, I found that unacceptable due to the large hole in its centre that was obviously the mode of entry during the earlier riots. The room next door had an undamaged working door so I moved in and collapsed on the single bed and slept through the night in an exhausted stupor.

The next three days were spent acclimatising to life in Freetown. I met many Sierra Leoneans and expatriates and all were generous with their hospitality. I also learned that the state of unrest in the country had not ended and that it was still unsafe to travel out of Freetown. I heard many horror stories concerning the riots – local people had been killed and Europeans, Zimbabweans and Nigerians on campus had been beaten by marauding mobs, high on the drugs jamba and alcohol. It was with some trepidation that I went to bed every night, after checking more than once the locks on the door and on the French windows, above which the moonlight filtered through glass louvres.

My fourth night in Sierra Leone was a particularly nervous one. I had difficulty getting to sleep because the next morning, at 9.00 a.m., I was due to give the first lecture of my career, an introduction to micropalaeontology for third-year undergraduates. I undressed and laid naked under the mosquito netting in my own personal sauna. My windows and doors were all secured but the lack of air-conditioning resulted in a room so hot and humid that I half expected rain clouds to form beneath the ceiling.

I am deaf in one ear and always sleep with my good ear down, oblivious to the world around me. That night, for whatever reason, perhaps the heat, I turned over in my sleep and a noise insinuated itself into my consciousness. I looked up into the room and my eyes gradually adjusted to the darkness. The scraping noise continued and I began to realise that someone was trying to break into my room. I struggled to get out of my bed, of course entangling myself in the mosquito net. Once extricated, my eyes focused on the glass louvres above the French windows. But the glass wasn't there. Instead, the space was filled by the heads and shoulders of three large Sierra Leoneans, who were obviously manoeuvring themselves to jump down into the room.

What was I to do? I did what any Englishman would do – I put on my underwear. I was not going to die naked! I yelled (and probably screamed) but the three men did not back off. I had never before hit anyone in my life but I decided this was the time to start. I reached into the corner by my bed for the solid walking stick that I used to keep the dogs at bay whilst walking around the campus after dark. I swung the stick and hit the intruders hard. But they did not retreat. I wondered why they didn't cry out in pain, and then I realised why. They had knives between their teeth!

'Run! Run, Culver and wake the night watchmen! Don't worry about all of your worldly goods, ripe for the picking, still packed in your suitcase.' These were the thoughts that flashed through my mind. I strode to the door and struggled with the handle until I realised that the door was, of course, still locked. I fumbled for my keys in the darkness, yelling at the top of my lungs, the skin on the back of my neck crawling, expecting at any moment a blow from behind. Finally, I managed to unlock the door and, without looking back, I ran to the front door of the Guest House to alert the night watchmen. The door was locked from the outside; the night watchmen were nowhere to be seen – I was trapped.

I ran down the corridor knocking on doors, trying to wake anyone in the adjacent rooms. No answer. I ran up the stairs and attacked the doors on the first and second floors. Again, no answer. Back down the stairs once more to try to rouse the misnamed night watchmen. On the way, I beat once more on a first-floor door. Suddenly, it burst open and my life once more flashed before my eyes as a baseball bat wielded by a large blond man whistled past my nose. Before the second swing came I managed to blurt out something, I don't know what, that stopped the attack. I found out later that the man, a visiting Fulbright Scholar from Pennsylvania, had been barricaded in his room with his wife and baby when the Guest House had been ransacked two or three weeks previously. He had obviously been determined to fight back on this occasion – but his 'attacker' was a pale, skinny, scared to his non-existent boots, half-naked Englishman.

The American's bulging, wide-open eyes narrowed as his adrenalin rush subsided and he realised I was no threat. I quickly explained what had happened and he pulled me into the room. On hearing that all of my belongings were still in my bedroom he persuaded me (unbelievably, with hindsight) to return to retrieve my suitcase. My suggestion that he come with me was politely declined but I did manage to convince him to lend me his bat. I crept down to my room and peered through the door. The room was empty and through the French windows I could see the night watchmen gesticulating and shouting towards the dense bush that surrounded the Guest House. My yells must have wakened them but I was *not* going to sleep again on the ground floor. I went back with my case to my new American acquaintance's room and begged shelter. Somewhat incongruously I was formally introduced to his wife who was dressed considerably more modestly than I was.

We were putting together a makeshift bed of blankets on the bare concrete floor when we heard the squeal of tyres and the screech of brakes. The police had arrived. I looked out of the window to see several officers baling out of two Land Rovers and, amidst much shouting, guns were drawn. On the command of the officer in charge shots were fired. At first I assumed that the guns were aimed into the bush where the three teefmen (local parlance for thieves) had disappeared. But no, the guns were pointed upwards and with a sudden, awful sick realisation, my brain told me that the bullets were striking the building only feet away from the window through which I was sticking my head. I dropped to the floor to join my American friends and wished with all my heart that I knew where the toilet was located.

Morning came before sleep. My body was a mass of angry red itching welts where mosquitoes had struck through the thin cotton blanket that I had wrapped around myself as I lay in a quivering foetal position. In the dark depths of the night I had decided to catch the next plane back to good old England but as the sky lightened so did my spirits. I thanked my hospitable friends profusely and then spent the next hour arguing with the landlady. I insisted on a new room on the top floor of the Guest House and, finally, my intransigence won the day. But this delay resulted in my late arrival in the Department of Geology. I had failed to turn up in time for my first-ever lecture. I hurried through a group of students huddled in the corridor complaining loudly about the non-arrival of their new teacher. I begged the Head of Department's forgiveness, which he immediately gave. Apparently, he was a relative of the Guest House landlady and word had already filtered back to him about my nocturnal excitement. But to my horrified surprise he instructed me to give the next lecture, due to begin in 15 minutes. The subject was West African geology of which I knew nothing. My objections fell on deaf ears and I was given a set of someone's illegible lecture notes to read to the class.

I stumbled off to the lecture room and stood in front of 20 expectant African faces. Somehow I struggled through my lecture as the students stared at me, not one of them taking a single note. They filed out quietly at the end and I was left standing there, a mental wreck. I staggered off to the Guest House and placed myself at the bar where I sat motionless until it opened two hours later. I did not leave until the bar closed for the day but by that time I had realised that I had survived 24 hours of hell that would never be repeated. I had given my first lecture in circumstances that I could never have imagined and I decided that, after surviving this, lecturing before any sort of audience would never hold any fear for me – I had looked through the gates of lecturers' hell and survived.

Some weeks later it became 'safe' once more to travel up-country. During the unsafe period many travellers were stopped at improvised road blocks, forced from their vehicles and robbed by thugs armed with rusty machetes. These unfortunate travellers were left standing naked at the roadside waiting, hopefully, for a more friendly reception from the next people they encountered along the road. Now that it was 'safe' the persons manning the roadblocks simply demanded a toll that would guarantee safe passage.

I now determined to take my students on their first geological field trip beyond the Freetown complex. I had found it very difficult to communicate the idea of stratified sedimentary rocks because none of my students had ever seen any. Perusal of the geological map of Sierra Leone showed me that I needed to drive 130 kilometres inland to examine the Rokel River Group. These sedimentary rocks were mapped as spanning the boundary between the Precambrian and Cambrian periods (the interval in Earth's history when animals with hard parts appeared in the fossil record), and so I intended to look for fossils in the upper parts of the sedimentary sequence.

My students, with few exceptions, were Creoles, city dwellers from Freetown. On the morning of the field trip they arrived, both male and female, dressed to the nines in the fashion of the day, flared trousers and platform shoes. In stark contrast I wore shorts and walking boots. There was a reason, of course, for this difference. The reason was driver ants. If one has the misfortune of standing on a column of migrating driver ants, the fighters, armed with large pincers, rush to attack the intruder – in this case human feet and legs. They bite viciously and hang on. They cannot be brushed off but must be picked off one by one. This is a painful process when many ants are involved. But, with shorts, at least the ants can be seen. My first field trip up-country was enlivened by several of my students (much to the amusement of their peers) rapidly ripping their trousers off before the driver ants that they had stepped in reached particularly important parts of the anatomy.

On an overnight trip several months later, we camped in tents on a sandy beach beside the Little Scarcies River. A British colleague from Fourah Bay College came to stay with us that night. Geoffrey had lived in Sierra Leone for some 20 years and insisted that he had no need for a tent. He arranged his camp bed under a tree, suspended his mosquito net and settled down to sleep. I zipped myself securely inside my tent. Sometime during the night we were woken by piercing screams of anguish. Ten geologists emerged from their tents, hammers at the ready, and were confronted by the awful sight of our visitor jumping and jerking in the moonlight like a demented

Figure 6.1 My students and me (about 1978) outside the Department of Geology building at Fourah Bay College.

marionette. While he had slept, a colony of driver ants had begun to migrate across the beach. On encountering our friend's camp bed the ants simply climbed up one leg of the bed across Geoffrey's chest and down the other camp bed leg to continue across the beach. How long this had continued, we will never know, but at some point Geoffrey had moved in his sleep and the ants had attacked. Hundreds of them scurried over his body and sunk their jaws into whatever flesh was available. Poor Geoffrey was in such pain we had to hold him down while each ant was individually picked off his body.

But, believe it or not, come morning our visitor considered himself a very lucky man. As he picked up his pillow a large centipede, over 20 centimetres long, fell to the floor. He had slept the night with this poisonous monster just inches from his face!

Driver ants are not all bad. A field trip to the heavily forested southern quarter of Sierra Leone showed me the imaginative use to which the local population put these vicious little beasts. Whilst walking down a jungle path, one of my party tripped over a root and stumbled into a tree trunk covered with sharp spines. He opened a 5-centimetre-long gash on his upper arm. I scrambled for my first-aid kit but after cleaning the wound it was clear that it would be difficult to keep the skin together. Our local guide solved the problem. He disappeared into the undergrowth and returned several minutes later with a matchbox filled with live driver ants. He took one at a time and held the head against the cut. The ant bit with one mandible on either side of the gash. Our guide then twisted the ant's body and separated it from the head, leaving a perfect suture. He repeated this exercise until some 20 ant heads were arranged along the cut. He explained that in a few days the muscles that held the ant heads impaled in the skin would decompose and the jaws would loosen and fall out. Indeed, this happened and my student was left with a neatly healed scar to show to his family on his return to Freetown.

In addition to my teaching duties at Fourah Bay College I was also required to undertake palaeontological research. This I did gladly because scientific publications are a prerequisite for career advancement. But my research did not progress without a hitch or two. One of the reasons I was offered my job in Sierra Leone was that I was trained in biostratigraphy (the subdivision of sedimentary rocks based on their fossil content). The microfossils that were my speciality, foraminifera (single-celled marine organisms with tiny shells), are small and abundant and so are particularly useful for subdividing the sedimentary rocks encountered in borehole cores.

Before travelling to Sierra Leone I had been informed that several cores through the Cenozoic Bullom Group that underlies the Sierra Leone coastal plain were stored at the Geological Survey offices in Freetown. It was my job to set up a biostratigraphic subdivision of these core materials that could act as a framework for subdividing sediments of a similar age that would be encountered in boreholes planned by petroleum companies in the near future. I was disappointed to find that nobody at the Survey knew what had happened to the cores. They had lent them to a company but they could not remember the company's name. The cores were lost.

I did not want my research career to end there and so I planned to collect my own foraminifera from a section of the Bullom Group exposed in cliffs near Lungi Airport. After many hours of fieldwork and laboratory work I came to the unfortunate realisation that any foraminifera that may have been present in the sediments I collected from the cliffs had their calcareous shells leached away by

Figure 6.2 Scanning electron micrograph of the foraminifer *Spiroloculina* from the coastal waters of the Freetown Peninsula. The specimen is about 0.5 mm across.

acidic water percolating through the sediments during thousands of years of rainy seasons.

So what could I do? I decided on a new tack. Instead of trying to find fossil foraminifera to study, I would work on living foraminifera. The idea was to understand what controlled the distribution of modern foraminifera so that I could use that knowledge to help interpret the distribution of fossil foraminifera. These tiny animals live in all marine environments and so, being a non-swimmer, I chose the shallowest water available. I paddled into the sheltered waters of several small bays along the Freetown Peninsula and found that many foraminifera live attached to seaweeds. So I made my plan. I would order my own alcohol from England (my department had no budget for chemicals) and use it to preserve the organisms in my samples prior to staining them with rose Bengal, a protein-specific stain that would show which individuals were alive and which were dead.

I waited four months for the alcohol to arrive. The day it was received on campus it was stolen, evidently by someone who required a free drink! I then did what I should have done six months earlier. I went to town and bought some cheap gin and used that to preserve my live foraminifera. Now I was on a roll. The stain worked and I was ready to extract my specimens from the sandy sediment that comprised the sample. This is done by scattering the sediment thinly onto a small tray. The sediment is then examined through a binocular microscope and individual foraminifera are picked out by lightly touching them with the dampened sable bristles of a very fine paintbrush.

I looked through my microscope at the sediment scattered on a picking tray. I was overjoyed to see that the sample was full of beautiful (to my eyes) recently live foraminifera, their protoplasm stained a bright shade of pink by the rose Bengal. After months of setbacks I was now ready to pick the foraminifera, transfer them to specimen slides and to begin collection of data on the distribution of live foraminifera on the various types of seaweed. I reached into my desk drawer for one of the several fine paintbrushes I thoughtfully had brought with me to Sierra Leone. I dampened the business end with a little water and looked down the microscope so I could touch the bristles against one of the pink foraminifera. The metal end of the brush ploughed through the sediment: there were no bristles. In a panic I grabbed the remaining five brushes: again, there was not one bristle between them. The cockroaches that inhabited my desk had tired of eating my lecture notes and had supplemented their diets with the sable hairs of my picking brushes. It was at that moment I realised that the gods might have been telling me something about my desire to be a professional palaeontologist. The remaining gin was put to good use.

Figure 6.3 My luxuriously furnished living room in Freetown – home for two years.

Insects were a recurring theme during my time in Sierra Leone. The cockroaches competed with me to see who would eat my food (vegetables from street vendors, packaged food from the supermarket) the most rapidly. They made a further point by eating the insulation in my oven. Whenever I cooked (I learned to make my own bread), the oven gave off the heat of a blast furnace. At least it was so hot that it kept the rats that ate the cockroaches out of my kitchen during my rather infrequent culinary efforts.

But cockroaches were not my only bane. One evening I was sitting alone in the living room of the concrete bunker that I called home. I was reading yet another Thomas Hardy novel (living on my own, my choice of novel was based on their length – the longer the better to offset the loneliness). My barred windows were open in the hope that a cool breeze might at some point struggle into the room. I was aware of an insect flying in through the window but took no notice – it was probably another moth that would become the meal of one of the several geckos that shared my home with me. I lived on the floor, they lived on the ceiling. The insect landed on my neck. The sudden pain made me reach up to brush it off and the burning sensation instantly transferred to my hand. The insect landed on the floor and I stamped on it, quite stupidly, with my bare foot. The pain in my foot had me hopping madly to the bathroom where I jumped fully clothed into the bathtub which was full of cold water (mains water only ran for a couple of hours in the morning so it was necessary to store water for evening ablutions).

The burning sensation slowly subsided and I surveyed the damage. My hand, foot and neck had ugly brown burns that begged to be scratched. The next day I took the crushed carcass of the insect to the Zoology Department. I learned that I had been the victim of a blister beetle, a close relative of the insect whose bodies are ground up to make the legendary aphrodisiac powder, Spanish Fly. This experience made me think that soft music, a glass of wine and a log fire might be a preferable alternative.

I spent 17 months in Sierra Leone and, sad to say, I never found a fossil. Most of the country is composed of Precambrian metamorphosed rocks, some 2700 million years old. The only sedimentary rocks in the country were the Bullom Group, the Rokel River Group and the Saionia Scarp Group, the first leached of fossils and the last two composed of ancient glacial deposits, unfavourable for the preservation of fossils. I did research on all of these rocks but my discovery of fossils in West Africa had to wait until 1984 when I returned with a student and explored the eastern parts of Senegal. Even then the fossils I found were so small (several would fit on a pinhead) that I hadn't realised I had found them until the samples were processed back in the laboratory several months later.

Not all of us palaeontologists get the chance to be on exciting and romantic expeditions such as a search for dinosaurs in the Gobi Desert. But I love my ugly little Senegalese fossils just the same. And looking back on my time in West Africa, I realise I love that part of the world as well. The friendliness of the people, the beauty of the countryside and the many incidents that occurred made my time there

Figure 6.4 Two broken spicules of the heteractinid sponge *Eiffelia araniformis* (Missarzhevsky) from south-eastern Senegal. Each specimen is about 200 micrometres across.

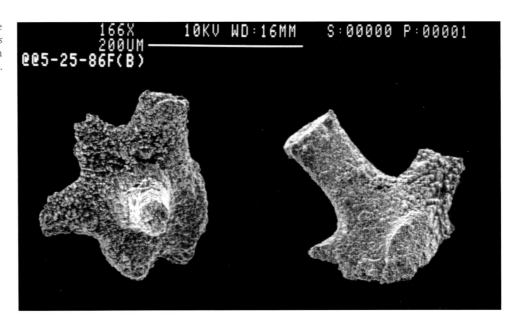

Figure 6.5 The village of Bumban in the Gbenge
Hills, Sierra Leone.

memorable. I have since lived in several different cities on two more continents. But no matter where I go I refuse to live in a house with louvred windows, I will never again read a Thomas Hardy novel, but I will listen to soft music, drink a glass of wine and sit in front of a log fire.

Chapter seven

Two passages to India

Paul D. Taylor

Geology, looking further than religion, knows of a time when neither the river nor the Himalayas that nourish it existed, and an ocean flowed over the holy places of Hindustan.

A Passage to India, **E. M. Forster**, 1924

Some people visit India to find themselves. I went there to find fossils – more specifically, fossils of bryozoans, a poorly known group of aquatic colonial animals. My two trips to India in 1991 and 1993 were to collect examples of bryozoans from rocks about 70 to 100 million years old. I count myself extremely fortunate in my career as a palaeontologist to have visited many different countries in the quest for fossil bryozoans. None, however, has had quite the impact of India. Even my extreme obsession for fossils failed to immunise me against the 'Indian experience.' I felt quite overwhelmed by this magnificent country and struggled to comprehend the extraordinary complexity and uniqueness of Indian life. This is an account of two visits to India that I have patched together from diaries, field notebooks, photographs and fading memories.

The north of England expression 'a bit parky' aptly described New Delhi at dawn on a late January morning in 1991. You could almost believe that the damp mist was due to the nocturnal exhalations of the capital city's quarter of a million inhabitants in the chilled air. As we drove from the airport a herd of military camels was being exercised at the side of the road in readiness for the forthcoming National Day military parade, adding to the exotic atmosphere.

I was in Delhi only very briefly to meet representatives of the Indian National Science Academy (INSA), who were sponsoring my trip as part of an exchange agreement with the Royal Society. INSA gave me a first taste of legendary Indian bureaucracy. The office to which I was led was a vista of constant comings and goings by numerous deferential clerks, part of an elaborate hierarchy of employees occupied in the movement and endorsement of countless pieces of paper. Eventually I was presented with an envelope bulging with rupees to cover my expenses while in India.

Why did I go to India? In the late 1980s an Indian geologist,

Dr C. Rajshekhar, visited The Natural History Museum (NHM) in London. He brought with him some specimens of a rock, called the Coralline Limestone, from north-central India to ask my opinion on the bryozoans. These could be seen in great abundance when the rock was viewed through a hand lens. The Coralline Limestone, which is of mid-Cretaceous age, had been given its misleading name by W. T. Blanford in 1869 before the correct identity of the dominant fossils was understood. One particular bryozoan stood out because of its sheer abundance and, to my experienced eye, unusual structure. This bryozoan had previously been described under a variety of different names by two generations of Indian palaeontologists but its true identity had not been appreciated, as I was to discover subsequently.

Here I will digress to explain what bryozoans are. Bryozoans are a taxonomic group (phylum) of colonial invertebrate animals (that is, without backbones) that live in abundance in the sea today and also occasionally in freshwater lakes. They are, however, seldom noticed and few people without biological training are likely to know of their existence. Bryozoans have no widely used popular name, although they are sometimes called 'moss animals,' 'sea mats' or 'lace corals,' and they do not figure in the diets of even the most adventurous gourmets or bizarre culinary cultures. They are often mistaken for corals or seaweeds. For example, one of the commonest objects found among the flotsam and jetsam of British beaches is a crinkly, pale-brown organism, sometimes called a 'hornwrack,' which looks like a bleached seaweed but which is in reality a bryozoan belonging to the genus *Flustra.*

Colonies of *Flustra* cast ashore during storms represent the tip of the iceberg of the rich and diverse bryozoan faunas inhabiting the sea-bed further offshore, not only in Britain but worldwide. For instance, over 900 bryozoan species live around New Zealand alone where they are key constituents of sea-bed communities, providing food and shelter for other animals, including juveniles of commercial fish.

Bryozoans vary in shape from thin encrusting sheets to bushy, tree-like colonies and perforated fronds. All bryozoan species build colonies containing anything between a few and many thousands of individual modules called zooids, each a millimetre or less in size. The zooids are furnished with a cone-shaped tentacle crown used for gathering planktonic food particles – microscopic plants and animals drifting in the sea . In common with many such suspension feeders, bryozoan colonies do not usually need to move because they are constantly provisioned with fresh supplies of food by the ever-flowing, plankton-laden water around them.

Fortunately for the palaeontologist nearly all bryozoan species have skeletons made of the calcium carbonate minerals calcite and aragonite. Both of these minerals, but calcite in particular, have a good potential for surviving as fossils. After the soft parts of the bryozoan have decayed, the skeletons remain and may be fossilised if covered by sediment on the sea-bed. Occasionally in the geological past, bryozoan skeletons were so common that their burial produced bryozoan limestones, like the Coralline Limestone.

Although bryozoans have a fossil history stretching back nearly 500 million years into the Ordovician period, they have been acutely neglected by palaeontologists. This neglect is partly because they are difficult to study and identify without either preparing thin sections made by grinding a slice of rock to a thickness of one-fiftieth of a millimetre until it becomes translucent, or by examining the surfaces of specimens using a high-powered scanning electron microscope. Not surprisingly, knowledge of Indian fossil bryozoans is exceedingly poor.

One of the most interesting times in bryozoan evolution occurred during the Cretaceous period. Then, the dominant group of bryozoans (cheilostomes) found today began to diversify and increase in numbers, gradually replacing a more ancient group (cyclostomes) as the most abundant and diverse bryozoans. Our knowledge of this extremely critical transitional period in bryozoan evolution is based almost entirely on evidence from European fossils. This geographical bias may in part be due to a real concentration of bryozoans in the European region, but it also reflects a lack of research on Cretaceous bryozoans elsewhere, India included.

Prompted by the unusual bryozoans I had seen in the thin sections of Coralline Limestone, and by my interest in obtaining a more complete understanding of global changes in bryozoans during Cretaceous times, I decided that a visit to India was essential, and I was delighted when the Royal Society approved my application for funding.

On the evening after my arrival in Delhi I flew to Pune, the starting point for this and my subsequent (1993) fieldwork. Pune, often spelt Poona in older atlases, is about 190 kilometres south-east of Bombay. It was known as a refuge from the worst of the monsoons for the Bombay administration in the days of the Raj. The city lies on the Deccan Plateau, 570 metres above sea-level. Quite rightly, it is generally deemed to be one of the pleasantest of Indian cities, without the excesses of climate and poverty that characterise many other centres of population.

I was given the warmest of imaginable welcomes by my principal host Ramesh Badve, who greeted me at Pune Airport with three colleagues from the Agharker Research Institute (ARI). The five of us piled into one of the ubiquitous Ambassador cars used by all such organisations and headed off towards the peaceful setting of the ARI. I recall this journey for giving me a first taste of the 'real' India, an India where the streets heave with people and animals: promenading young men

Figure 7.1 Khandala Ghat in Maharashtra. Piles of lava flows, erupted during the late Cretaceous as the Deccan Traps, have been eroded to give the distinctive stepped landscape of the Deccan Plateau.

sporting moustaches and wearing open-necked white shirts; families of four miraculously squeezed onto a motor-scooter, the women's saris drifting perilously close to the wheels; auto-rickshaws dodging between lorries and buses; grubbing pigs, most with the hairs on their backs shorn to make paint brushes; foraging water buffalo; and dogs with udders but cows without. Alternating smells of food before and after its passage through digestive tracts, and snatches of Indian film music punctuated by car horns, completed this full frontal assault on the senses.

After a few days spent acclimatising to the heat, and delivering a lecture at the ARI for which a bouquet of flowers was an unexpected reward, we set off for the Narmada Valley, about 480 kilometres to the north of Pune. For the journey we had the service of a Mahindra jeep. This is not a large vehicle, so you can imagine my surprise when I found out that it was to carry our excellent driver Ramdas, Ramesh, Head of the Department of Geology and Palaeontology at the ARI, his colleague 'Raj' (the same Dr C. Rajshekhar who had first shown me bryozoans from the Coralline Limestone), a student field assistant called Suresh, and me, together with all of our personal baggage and geological collecting gear, plus two large cans for diesel. The diesel cans were needed because of a shortage of fuel brought about by the Gulf War; subsequently, I felt very uneasy when, given a veneer of legitimacy by our government number plates, we jumped to the front of long queues of tractors at the few filling stations still provisioned with stocks of fuel.

It took the best part of three days to reach Manawar, our main base for fieldwork in the Narmada Valley. Any frustration at the length of time we spent on the road in the hot and noisy jeep was more than made up for by the fascination of the journey. The routine sights of rural India were all novel to me: bullock carts transporting farm produce; lines of women carrying on their heads brass pots full water from well to village; schoolgirls turned out in smart uniforms and boasting immaculately plaited long hair; roadside banyan trees marked with orange and white hoops; and richly decorated, overladen lorries with tailgates bearing the ubiquitous message 'Horn Please O.K.', a request for overtaking drivers to hoot a warning of their presence. Whereas it is rare nowadays in Britain to see any sign of human activity in fields adjacent to roads, in India every field seemed to contain people tending the animals or cultivating the surprisingly large range of bountiful crops made possible by irrigation.

Geologically, most of the route took us over basalts of the so-called Deccan Traps. The name 'Trap' is derived from a Scandinavian word and refers to the step-like appearance of the landscape produced by these igneous rocks. The Traps were formed towards the very end of the Cretaceous period when a huge quantity of basaltic lava poured out in a series of flows over the ancient Indian shield. Nothing of this magnitude or character has ever been witnessed in historic times. Today, the Deccan Traps cover 520 000 square kilometres and attain a maximum thickness estimated to be over 2000 metres in the Bombay area. Some geologists believe that the global climatic effects caused by such immense outpourings of lava were responsible for the mass extinction of dinosaurs and other animals about 65 million years ago – a counter theory to the asteroid impact hypothesis, which is often invoked to explain

Figure 7.2 The entrance to the Kailasa Temple, cave number 16 at Ellora. The temple is carved into solid basalt of the Deccan Traps.

the mass extinction of plants and animals, dinosaurs included, at the end of the Cretaceous.

It is fitting that these products of an enormous physical force of nature should have served as the raw materials for a human endeavour of comparable magnitude. The Ellora Caves in northern Maharashtra must rank among the most remarkable religious excavations anywhere in the world. Between the fifth and eleventh centuries AD, 34 caves were cut into the basalt on the hillside at Ellora. The changing 'architecture' of the caves chronicles a transition from Buddhism to Hinduism to Jainism. The beautiful rock carvings are still remarkably fresh, a testimony to the excellent weathering qualities of the basalt rock. But it is the scale of the excavations that is most breathtaking. The digging of cave sixteen – the Kailasa Temple – alone involved the removal of 85 000 cubic metres of basalt, to form a courtyard 30 metres by 45 metres, with a tower, cupola and gateway left standing in the unexcavated rock, an achievement made even more stupendous by the extreme hardness of the rock and lack of any mechanical digging tools.

Our first sight of the Narmada River was at Onkareshwar. This island in the river is a place of pilgrimage where holy men smear their bodies with ash, while monkeys scamper along the walls of the temples engaged in more sensible activities. There is no mistaking the temples at Onkareshwar. They are large and magnificent. The same cannot be said for some of the smaller temples dotted around the countryside, as I was to find out when attacking a pile of rocks with my hammer, failing to spot the dash of orange paint that signified the rocks were a temple. The villagers of Agarwara were forgiving and the ill-fate that I feared might be a consequence of my careless blunder never materialised.

Securing accommodation did, however, prove to be a source of anxiety throughout our trip. Nothing certain had been arranged in advance and all too often we found ourselves arriving in a town late at night, long after the best beds had been taken. This meant that we were forced to book into hotels where the addition of a 'luxury tax' to the bill seemed like an ill-judged irony. Apart from hotels, we also stayed in three different Government Guest Houses. These large bungalows are a legacy of the days of the British Raj and are still maintained for the convenience of government officials visiting the regions. The key to our admittance as bona-fide guests turned out to be the government number plates on the Mahindra jeep. Once the custodian of the Guest House had been persuaded to acknowledge this as an emblem of our 'officialdom,' we would make ourselves as comfortable as possible in the basic accommodation, and send Suresh and Ramdas out for chapattis, dahl-fry and 'Kingfisher Lager' or, where that was unavailable, 'Thumbs Up,' an Indian cola with more than a distinct hint of curry. Our evenings were spent on the verandas of the guest houses, eating, drinking and discussing our finds of the day and plans for the following day's fieldwork.

The best Guest House was at Manawar, a bustling settlement on the Man River, a tributary of the mighty Narmada. The lives of the inhabitants of Manawar range from the ordinary to the exotic. At the ordinary end of the spectrum are impoverished men scraping a living by selling as firewood long branches gathered from the thin forests above the town. The exotic was represented by a podgy transvestite eunuch who proudly strutted through the streets.

Figure 7.3 An inquisitive tribesman seated with
Ramesh Badve near Manawar.

Nearby Bagh Town had a very different feel from Manawar, with the
claustrophobically narrow main street giving it something of a medieval atmosphere.
Kneeling traders ply their cloth and other wares from the aprons of open shop fronts
that double as family homes. Bagh is well known geologically as the type locality of
the Bagh Group, a thin sequence of mid-Cretaceous sedimentary rocks sandwiched
between underlying rocks of the Precambrian or younger Gondwana Group and
overlying lavas of the Deccan Traps. The Bagh Group is exposed as a series of
disconnected outcrops, mostly in the valleys cut by the northern tributaries of the
Narmada River. Four main subdivisions are recognised: Nimar Sandstone, overlain
by Nodular Limestone, Deola-Chirakhan Marl and, finally, Coralline Limestone. We
found the richest fossil faunas in the last two units, with the softer Deola-Chirakhan
Marl proving best for collecting purposes and yielding abundant bryozoans as well as
many brachiopods (lamp shells), bivalves (cockles and their relatives) and echinoids
(sea urchins). Perhaps the most striking fossils were the echinoids that weather out in
great numbers at some localities. A local tribesman who crouched down beside us at

one locality already knew of their presence in the rocks and even had a name for them – *punchu khada* – meaning 'five-stone', an allusion to the five radiating 'ambulacral' rays on the surfaces of the fossils that mark the position of the tube feet in the living animal.

Ramesh had only recently recovered from major heart surgery but he excelled in the field. Reaching the fossil localities was not always easy as we had no decent base maps, the roads rapidly deteriorated into deeply rutted cart tracks, and obtaining any help from the locals was hampered by language differences when we were in the states of Madhya Pradesh and Gujarat. I was thankful we were not here during the monsoon when the dusty roads must have deteriorated into impassable rivers of mud. Ramesh's knowledge of the area and his sheer persistence meant that we reached all the localities we planned to visit. Several consisted of small quarries worked by hand, men digging the rock with picks while women carried it in head-slings down the quarry face to awaiting lorries. The main use for the quarried limestone seemed to be as hardcore or in cement manufacture. In former times, however, the Coralline Limestone was worked more carefully and polished for use as a building stone in temples and palaces, as at Mandu the fort-capital of the Paramar rulers of Malwa. Geologist P. N. Bose in 1884 noted that the Coralline Limestone 'takes a fine polish, and the thick clusters of branching Bryozoa, of which it is largely made up, give it a most picturesque appearance.' The ruined temple at Deola is a smaller-scale testimony to the beauty and lasting quality of well-sculpted Coralline Limestone.

Figure 7.4 Sculpted Coralline Limestone figurines in the temple at Deola.

Politically, this was not a good time to be visiting the Narmada Valley because of unrest caused by the construction of the massive Narmada Dam that would force the relocation of many of the tribal inhabitants of the region. But we encountered no hint of hostility whatsoever from the people we met. The villagers, curious about our strange behaviour with hammers and plastic collecting bags, were always friendly. At Bilthama they brought us sweet tea to drink and watched while we collected fossils from the slopes beneath their village. To my surprise, some of the tribals carried bows and arrows, and not for hunting animals. Ramesh told me that these were seldom used, except when alcohol stoked high spirits during religious festivals. He also knew a story of arrows being fired at government officials attempting to promote family planning, not popular in a society where having a large family increases the probability of children being around to look after the parents when they become old and infirm.

Our return journey to Pune was along the west coast via Surat and Bombay, where we collected modern bryozoans encrusting shells on the muddy beach at Elephanta Island, famous for its caves with giant figures carved in Deccan basalt.

The significance of fossil bryozoans collected in the field seldom becomes apparent until cleaning, sorting and microscopic study have been completed back in the laboratory. And so it was with our material from the Bagh Group. The peculiar bryozoan that had first attracted my interest turned out to be a new genus, which Ramesh and I went on to name *Chiplonkarina* in honour of the late Professor G. W. Chiplonkar who had worked extensively on the fossils of the Bagh Group and other Indian Cretaceous rocks. Bryozoans are notorious for showing 'homeomorphy,' whereby different groups evolve similar colonial architectures independently, repeating a successful design for living. *Chiplonkarina* is an extreme example of such homeomorphy. Previous studies of this dominant Bagh Group bryozoan had always placed it among the cyclostome bryozoans because of its curved tubular zooids, which are characteristic of cyclostomes, in contrast with the box-shaped zooids generally typical of cheilostomes. Other features show quite conclusively, however, that *Chiplonkarina* is a cheilostome mimicking a cyclostome to an extraordinary degree. After our description of Indian *Chiplonkarina*, the genus was subsequently recognised in Cretaceous rocks in France, Germany, Turkmenistan and New Zealand, although nowhere does it occur in such abundance as in the Bagh Group.

My appetite for India and its fossil bryozoans had been whet and I resolved to pay a return visit with a view to sampling the rich Cretaceous bryozoan faunas known from South India. I was able to do this, under the sponsorship of the British Council, in 1993.

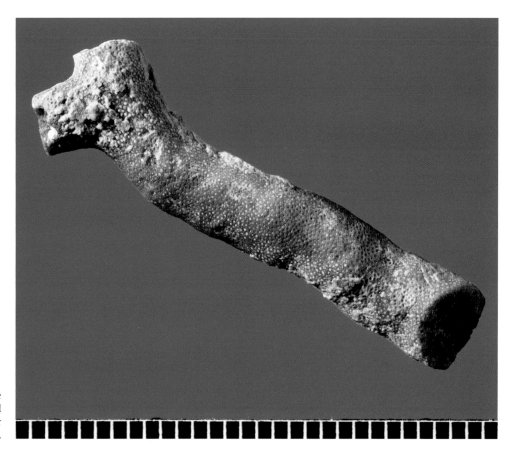

Figure 7.5 Branching fragment of a colony of the bryozoan *Chiplonkarina dimorphopora* collected from the Deola-Chirakhan Marl at Badia. Scale bar divisions are in millimetres.

In a place close to the border between the states of Karnataka and Tamil Nadu there is a very ordinary-looking wall bearing a sign painted in four languages: English, Hindi, Karnatakan and Tamil. The sign requests gentleman to refrain from relieving themselves on the wall, although not in such polite terms. My 1993 trip certainly taught me about the multitude of divisions that dissect this country: divisions of language, as illustrated trivially by the wall; divisions of religion, reflected in the racial riots that had recently wreaked havoc in Bombay, leaving some streets etched with acid burns; and divisions of caste that were brought into sharper focus by reading V. S. Naipaul's *India. A Million Mutinees Now* during my journey.

I also came to realise how differently palaeontology was approached in India relative to Western countries. Palaeontology is a massive subject with too few practitioners. Good communications and a spirit of co-operation between palaeontologists in Western countries generally ensures minimum duplication of effort by different groups of researchers, and when two or more groups work in similar areas they freely acknowledge and constructively criticise each other's work. Such is not always the case in India: a research group in one university may study fossils from a particular field area with no knowledge of the existence of 'rival' groups, or at least making no reference to their published results. In my opinion, this has greatly hindered progress in Indian palaeontology, as has the inclination to reiterate the findings of the pioneering geologists, who mapped and surveyed India during the nineteenth century, and reluctance to build upon this work with new results and interpretations. I wonder also whether divisions of caste are obstructive in that palaeontology requires its practitioners to combine the activities of a labouring caste – hammering at rocks and grubbing around for fossils in the field, as well as scrubbing them clean and handling them back in the laboratory – with those of a thinking caste. Many academics are Brahmins.

In a reprise of our 1991 field trip, four scientists piled into the back of the Mahindra jeep and set out from Pune for a three-day journey of over 1000 kilometres. Ramdas, again our driver, carried a portable radio with him and took delight in teasing me about the performance of the England cricketers, who at the time were also travelling through India and being soundly beaten by Kapil Dev's Indian eleven. My geological companions this time were Ramesh, his student Mohan Sonar, who was undertaking a doctoral study of fossil bryozoans of more recent times (the past 10 000 years), and Ajit Vartak, an ammonite specialist who taught in a small Pune college. Once again we headed out across the sun-baked Deccan Plateau, spending three gruelling days on perilous roads where overturned lorries were commonplace, few vehicles travelled with lights after sunset, and the pot-holed surface forced Ramdas into violent swerves, mirrored more alarmingly by the on-coming traffic. I was puzzled by the phrases 'We Two' and 'Our's One' painted on many of the lorries until Ramesh explained that they were slogans concerned with limiting the size of families.

The roadside banyan trees of the north are replaced by tamarind trees in the south, and our diet of dahl and chapatis by idli (cakes made from fermented ground rice) and samba (a watery curry incorporating pods from the same tamarind trees). Southern Indian food is far removed from the cuisine offered in many Indian restaurants in Britain. Although the 'real thing' is clearly wholesome, I'm afraid I found it less palatable than the Anglicised version and was inclined to skip lunch when we stopped at roadside cafes. These were invariably equipped with couches of rope strung across wooden frames for the diners to enjoy a siesta after eating. Instead I adopted a frugivorous diet of miniature bananas and sweet limes, which had the welcome side-effect of reducing my weight by almost a stone during the three weeks I spent in India.

Our journey from Pune took us south to an overnight stop at the Hotel Peacock, Hubli, where the cockroaches were a poor faunal substitute for the absentee ornamental birds from which the hotel had taken its name. From Hubli we drove to Bangladore, a city distinguished by wide boulevard-like approach roads, abundant greenery, proliferating high-tech industries, and the nearest thing to an atmosphere of prosperity I encountered in India. I think it was in Bangalore that I made an unwise purchase of some locally produced 'mineral water,' failing to notice the brown deposit at the bottom of the bottle. My discomfort on the following day's journey had just about subsided when we stopped at a tidy farming settlement of four small huts, the home of a 15-strong Tamil family known to Ramesh, and were offered cups of cool, slightly efflorescent milk from freshly harvested coconuts. Like virtually all the rural settlements we saw in India, piped water, electricity and sanitation were nowhere to be seen.

By now we were travelling over ancient igneous and metamorphic rocks of Precambrian age that form the foundation of the Indian shield. Until roughly 120 million years ago India was part of a southern supercontinent called Gondwana, joined onto Africa, South America, Australia and Antarctica. The name Gondwana actually derives from an ancient Indian tribe called the Gonds whose descendants still inhabit Madhya Pradesh. When Gondwana broke apart, India began a northward sprint (in geological terms), crossing the equator and eventually battering into Asia. Sir Edmund Hilary's moment of glory on Everest would not have been possible without this collision, which was responsible for the formation of the Himalayas.

Younger sedimentary rocks have been deposited on top of the Indian shield at various times in the geological past. These include an important series of rocks, the Gondwana Group, formed while India was still joined to Gondwana and ranging from late Carboniferous to early Cretaceous in age. The Gondwana Group contains substantial amounts of coal and is therefore of great economic importance to India. Plant fossils are common in the Gondwana Group, as we were to witness in a fireclay quarry at Therani. Later in the Cretaceous, after India had disengaged from Gondwana, the sea advanced periodically onto the edges of the Indian shield, depositing marine sediments often packed with fossils of animals that had lived on the sea-bed. Two of the most important sedimentary basins filled by such Cretaceous fossil-bearing rocks are the Narmada Basin in north-central India, which I visited during 1991, and the Cauvery Basin in south India, the target of our 1993 fieldwork.

The Cauvery Basin edges onto the Precambrian shield south of Madras, in the region between Pondicherry and Trichinopoly. There are three patchy Cretaceous outcrops, the largest of which, centred on the town of Ariyalur, was our goal. The Cretaceous deposits here are much thicker and were formed over a greater length of geological time than the Bagh Group of the Narmada Valley. Like the Bagh Group, however, marine fossils are abundant and well preserved. Around Ariyalur the Cretaceous is divided into three stratigraphical groups, called from the oldest to the youngest, Utatur, Trichinopoly and Ariyalur; a younger, fourth division – the Niniyur Group – is of early Tertiary age. A rich bryozoan fauna was known to exist in the Ariyalur Group, about 70 million years ago. These bryozoans were in acute need of revision having been neglected since the Austrian palaeontologist Ferdinand Stoliczka first described them back in 1873. Stoliczka's paper is notable for the fact that he referred to bryozoans as 'Ciliopoda,' a name of his own invention and now largely forgotten.

Through the kindness of Mr S. Lakshminarayanan, deputy general manager of Dalmia Cement Ltd, we were treated to accommodation in the Guest House of the company throughout the period of our fieldwork in Tamil Nadu. The Guest House chef cooked a particularly fine bhindi bhaji, and I even grew fond of his breakfast special of home-made cornflakes doused in hot milk, eaten while the electronic clock in the dining room celebrated the hour of eight o'clock with a brief medley of Christmas tunes. The Guest House occupied a small corner of the Dalmia Colony, a fenced and guarded settlement, complete with its own hospital, school, temple and workers' housing, which has grown up around the cement factory and associated quarries. For the railway buff there was even a narrow-gauge branch line with a steam engine bearing the ubiquitous Indian star on its smokebox door.

A series of large limestone quarries, worked by blasting for the manufacture of cement, pock-mark the flat landscape between Dalmiapuram and Ariyalur. Although none of the quarries has any underground workings, they are nevertheless usually referred to as mines, and almost all employ their own graduate geologists who were delighted to welcome us and to discuss fossils over a cup of sweet tea. The geologist at the Kallankurichi Limestone Mine proudly showed us a dinosaur egg resembling a stone cannonball that had been found in his quarry. But most of the fossils here and elsewhere consisted of a wide range of invertebrates that had inhabited the shallow Cretaceous sea, including sponges, corals, brachiopods, molluscs, echinoids, belemnites and ammonites (extinct cephalopods related to the nautilus) and, of course, bryozoans. Mohan, the bryozoan doctoral student, took great delight in bringing me his finds at regular intervals and proclaiming them to be 'pure Bryozoa.' A recently published monograph of the Ariyalur Group bryozoans by Asit Guha and D. Senthil Nathan of the Indian Institute of Technology in Kharagpur catalogues exactly 100 different species. This high diversity serves to underline the importance of these deposits for studies of Cretaceous bryozoans.

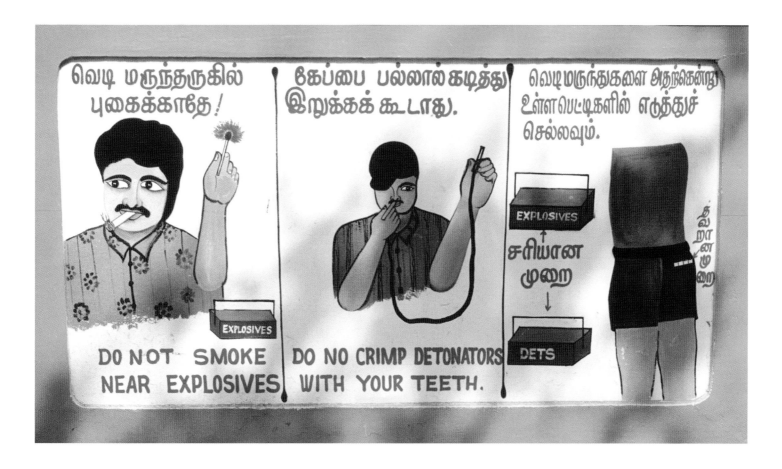

Figure 7.6 Hand-painted safety sign for the benefit of the workers at the Kallakuddi No. 2 Quarry.

Most of the bryozoan fossils in the Kallankurichi Limestone Formation division of the Ariyalur Group are thin encrustations on the surfaces of brachiopod and bivalve shells, while others are branch fragments broken from colonies that formerly resembled small bushes. Ramesh knew a wood carver in the village of Kallankurichi after which the limestone was named. I commissioned him to carve a tiger for my son James, paying a mere 200 rupees (about £5) for this together with a carved head of Ganesh the elephant-headed god.

While collecting fossils in one of the Kallankurichi Limestone quarries I accidentally disturbed a wasps' nest and was promptly stung twice, once on each hand. Not being one to keep such misfortunes to myself, I let out a yell. Ramesh immediately came over to me and asked for my hands. He rubbed the stings with his fingers and the pain disappeared instantly. Claims of faith-healing and the like seldom impress me, but I can venture no scientific explanation for this particular incident.

In the village giving its name to the older Utatur Group are two wooden chariots, one thought to be 200 years old and the other perhaps 800 years old. These

Figure 7.7 Cretaceous dinosaur egg from the small museum at the Kallankurichi Limestone Mine.

elaborately sculpted chariots ride on thick wooden wheels and are intended to be towed through the village during religious festivals and celebrations. The themes of the carvings are as imaginative as they are sexually explicit.

We were successful in discovering the first unequivocal bryozoans to be found in the Utatur Group. They consisted of small dome-shaped colonies less than a centimetre in size. Among them is a species belonging to an unusual extinct group of cyclostomes called melicerititids that had never been found outside Europe, thereby considerably extending the geographical range of these animals and proving that they spread south of the equator during Cretaceous times. The bryozoans occur in sediments that can be traced laterally into reefs full of corals, sponges and other fossils. Angular and jumbled blocks of reef limestone up to 5 metres high can be observed in one spectacular bed at the KVK No. 1 Quarry, possibly the result of Cretaceous earthquakes causing chunks of the reef to break off and tumble downslope into deeper water.

As in the Narmada Valley far away to the north, we encountered nothing but friendliness and generosity from the local people. One farmer and his wife gave me

Figure 7.8 The more ancient of the two religious chariots at Utatur.

lemons. Squeezed into my plastic bottles of Basleri mineral water, the lemons counteracted the taste of plastic that developed as the water reached high temperatures in the hot interior of the jeep. An old man tending an irrigation well went out of his way to present me with a bag of fragrant petals for no other reason than the fact that I was there. Generosity is so often inversely proportional to affluence. I was surprised by some of the styles of personal adornment fashionable among the Tamils. Women with yellow, turmeric-powdered faces were a common sight. Even more bizarre to my eyes were men wearing red or purple nail varnish.

One of the remarkable features of many villages in this part of India are the statues erected to guard the villagers against evil spirits. Modern examples may consist of painted life-size figures of soldiers with guns, but the guardians more traditionally take the form of horses. These are occasionally 5 metres or more high. Other architectural gems of Tamil Nadu are the Dravidian style temples with their steep terraced roofs festooned with myriad gods and other figures. The exterior of the

Figure 7.9 Guardian horses constructed in the village of Poyyudanallur to ward off evil spirits. Mohan Sonar stands beneath the larger, unpainted horse.

modest temple in Ariyalur is distinguished by brightly painted figurines. At Tanjore (Thanjavur), having failed to locate a small outcrop of Cretaceous rocks, we visited the enormous Brahadeeswarar Temple said to be the finest in India. A block of granite about 80 tons in weight rests on top of a 60-metre-high tower. One thousand years ago the stone was hauled there up a temporary incline nearly 6.5 kilometres long, in a feat reminiscent of the building of the Egyptian pyramids.

My departure from India in 1993 came suddenly. Word reached us at the Dalmia Colony that the engineers working for Air India, with whom my return flight was booked, had gone on strike causing drastic cancellations of services. Fortunately I managed to find a seat on a flight from Madras to Bombay, where, after a day's delay, the British Council booked me onto a British Airways flight home to London. Immediately after my return to Britain, the international news reported that a series of 13 terrorist bombs had killed 250 people in the Nariman Point area of Bombay, very close to where I had lunched with representatives of the British Council only two days before. And, as I write this chapter in June 1997, a major fire at the Brahadeeswarar

Figure 7.10 Painted figure from the temple in Ariyalur.

Temple in Tanjore, apparently caused by a stray firework, has caused almost 50 deaths.

These tragedies illustrate the self-destructive side of India. Associated with the cultural diversity that makes India such a fascinating destination for the tourist are sectarian tensions that periodically threaten to blow the country apart. Geology teaches us how India as a continent has survived its traumatic collision with Asia. One can only hope that India as a nation will also survive its current cultural collisions.

Chapter eight

Digging for dragons in China

Angela Milner

We feel that it is necessary to consider the Yiping form as a new type of Sauropoda for which the name *Mamenchisaurus constructus* is proposed. The generic name refers to the locality of the find and the species name indicates the find was made as the result of the constructional work so extensively made in our country.

'On a new sauropod from Yiping, Szechuan, China', Young Chung-Chien, *Scientica Sinica*, 1954

Sichuan, a Chinese word meaning 'four rivers converging onto a plain,' is a vast sedimentary basin in south-central China. Rocks laid down in lakes and rivers between about 245 and 130 million years ago record a brief but continuous episode in Earth's history, from the late Triassic to the boundary between the Jurassic and Cretaceous periods. The beds, more than 3000 metres thick, consist predominantly of purplish red mudstones, collectively called 'The Red Basin of Chongqing.' From them have come rich and varied remains of dinosaurs – the longest unbroken record of Jurassic dinosaurs from anywhere in the world.

That is the reason why, on a glorious, sunny autumn day in 1982, I found myself walking along the Great Wall of China, fulfilling an ambition I had held since childhood. It was a memorable beginning of a nine-week visit to China to take part in fieldwork in Sichuan Province. Alan Charig, then Head of the Reptile Section, and Ron Croucher, formerly senior preparator and Head of the Palaeontology Laboratory at The Natural History Museum (NHM), were my companions on an expedition to collect late Jurassic dinosaurs at the invitation of The Institute of Vertebrate Palaeontology and Palaeoanthropology (IVPP) in Beijing. Not only did we have the opportunity to co-operate with our Chinese colleagues in the field and live alongside the local people in rural northern Sichuan, but we had the unique privilege of visiting an area of China that was not normally open to foreigners.

Rolling down the runway at Gatwick on a September evening, bound for Hong Kong, convinced us that we were, finally, on our way after three years of uncertainties and postponements. The trip had been planned in 1979, when Alan visited China as a member of a four-strong delegation of British vertebrate palaeontologists under the auspices of the British Council's 'Academic Links with China' scheme. Our principal objectives were to work with the Chinese in the field, to

Figure 8.1 Dinosaur-bearing rocks in the Red Basin of Chongqing. The sloping beds of purplish red mudstone and overlying grey sandstone in the Tangjia He Valley in northeast Sichuan are part of the late Jurassic Shaximiao Formation.

exchange information and give demonstrations of our field techniques and to acquire, in return, a good replica skeleton for future exhibition in London. We originally hoped to travel in 1980 but it was eventually fixed for 1981. All was ready for an autumn departure, our field equipment and supplies had been freighted out and awaited our arrival in Beijing, but just three weeks before departure Sichuan suffered catastrophic flooding and severe disruption to communications. Gloom descended even before the inevitable telegram arrived. Our Chinese hosts had no option but to shelve the expedition because the area we were to visit, Wangcang County, had been one of the worst affected.

In 1982 we spent a week in Beijing at the IVPP; meeting colleagues and renewing old friendships, among them Li Jinling, a young researcher who had previously spent two years in London. At last we were briefed on the full details of the trip. We had left London with little idea of where we were going. The Chinese side of the field party included seven IVPP staff, dinosaur specialist Dong Zhiming, the leader, three preparators, two drivers, Li Jinling (whose services as interpreter were essential) and two representatives from the Chongqing Natural History Museum in Sichuan. Most of the team had gone on ahead to make final arrangements on site except for Li Jinling and Zhang Hong (the chief preparator at the IVVP), who were to travel with us. There was just one hitch. An important item of our field supplies – containers of polyurethane foam that were air-freighted out from London – had arrived late, several days after us, and too late to have been sent on ahead by road with the rest of our equipment. Polyurethane foam is used to make fast-setting, light-weight field cocoons to encase fossil bones, an alternative to traditional plaster of Paris. We had planned to introduce this technique during the forthcoming excavation. No problem, the containers could go on the train with us – or so we all thought.

Our hosts laid on some classic sightseeing besides the Great Wall – the Ming Tombs, Imperial Palace and much more. Visits to the magnificent collections of fossil vertebrates in the IVPP's private museum, Beijing Natural History Museum and Zhoukoudian, the site where Peking Man was discovered, reminded us that we were not just tourists. One thing we learned quickly was that the Chinese love a dinner party. A Cantonese banquet was thrown in our honour. We duly reciprocated the hospitality with a Beijing Duck Feast on our last night in the capital. Tradition demands that the guest of honour at a duck feast consumes a duck's head, neatly bisected, and wrapped in a pancake with spring onions and plum sauce (I was rather relieved that we were the hosts on this occasion). A very convivial evening was punctuated by toast upon toast as everyone relaxed and thought of yet another excuse to down yet another measure of mao tai. Mao tai, very strong firewater distilled from rice, has a somewhat unusual flavour and is best tossed back quickly, thereby minimising contact with one's tastebuds.

Feeling decidedly fragile the next morning, we struggled through the multitudes encamped in and around the railway station to board the Beijing–Chongqing express. We settled into our couchette compartment in the 'ruanwoche' or soft-seat coach. Such luxury was restricted to foreign travellers (we were the only ones on the train) and high-ranking Chinese, including military officers. Li Jinling shared our accommodation but Zhang Hong had to travel further down the train in a sardine-packed hard-seat coach. Our compartment was resplendent with embroidered linen, net curtains, lace tablecloth, the ubiquitous tea cups and vacuum flask, bonsai tree and a table lamp that refused to work until Ron turned it upside down in desperation. We discovered yet another facility as the train pulled out of the station – a loudspeaker that emitted raucous music and shrill speech. The volume control had not the slightest effect and, having failed to muffle the cacophony with coats and pillows, Alan ensured peace and quiet by disconnecting the wires. This was just as well, for music and commentary are a frequent accompaniment on China Railways, even at 3 a.m.

Our journey was punctuated by prolonged visits to the dining car where we consumed vast quantities of delicious food laid on especially for our little party of four. This particular train had been awarded a rosette for its Sichuanese cuisine and the chef certainly did himself justice; so did we, staggering back to our compartment in some discomfort. Discomfort increased for a different reason when we discovered, on visiting Zhang Hong further down the train, that the polyurethane was not with us. The Chinese authorities at Beijing Station had decided that it was too dangerous to be carried on a passenger train and flatly refused to allow the containers on board the baggage car.

The train journey lasted 33 hours with three route stages. The first was south and across the Huang Ho (Yellow River) to Luoyang in Henan Province, where diesel gave way to steam and nostalgia, especially from the older members of the party. The second leg, all in the dark unfortunately, was a long slow pull westwards up the Wei Ho valley, passing through the city of Xian, and another change of engine, electric this time, at Baoji. Dawn revealed the spectacular, mist-shrouded Micang Shan mountains as the train snaked its way slowly south down the narrow upper Jialing River valley on the Gansu–Shaanxi border, the single track following every curve and bend of the river.

There was much evidence of severe flood damage, the aftermath of the summer 1981 disaster that had led to the postponement of our trip. One of the main railway bridges over the Jialing had been destroyed, cutting communications for some months afterwards. As the train inched its way over the temporary replacement, a flimsy-looking structure of girders and planking, the main concrete columns of the old bridge could be seen, lying completely smashed, in the river bed some 15 metres below. Gangs of workers were engaged on repairs to the track and embankments along the valley, housed in temporary huts built within a hair's breadth of the rails and constructed of rush matting with a pile of earth to secure the roof. They were obviously due for a long stay, every square centimetre of land less than vertical had been planted up with cabbages and corn. The gangs were labouring in time-honoured Chinese fashion, employing the simplest of tools and earth-moving equipment – long-handled trowels and rush baskets. Sheer numbers made up for the lack of mechanisation.

We disembarked at stations along the valley to stretch our legs, much to the amazement of the locals who gathered round on the platform for a long stare or squatted on the rails, gazing inscrutably up at us through the train windows. Guangyuan, just south of the Shaanxi–Sichuan border, was our set-down point. There we were met, not only by Dong Zhiming and the rest of the IVPP team, but also by the mayor of Wangcang County, Li Wenzing, plus an entourage of local dignitaries who had made a two-hour journey just to welcome us. We were VIPs! The final leg of the journey, by minibus, took us east along the wide and fertile valley of the Dong He (East River) to the town of Wangcang. The driver proceeded at break-neck speed along a winding, bumpy road on sidelights alone, somehow avoiding dark shadowy two-wheeled figures with no lights at all. Bicycle lamps just do not feature in China. We turned into Wangcang hostel grounds 35 hours after leaving Beijing, almost too tired to notice the assembled crowds eager for their first glimpse of foreigners.

Hostels are official government residences where accommodation and meals are provided for local conferences, educational courses and visiting officials. Wangcang hostel, built within a walled compound, consisted of a dormitory block, kitchen and dining-room block, staff quarters and, set slightly away from the rest across a small garden with a lotus pond, a single-storey, distinguished visitors' building. This was filled entirely by the field party and we three were allocated identical, simply furnished rooms with concrete floors, whitewashed walls and magnificent four-poster beds. Alan's quarters had been occupied not long before by no less than the then Chinese Prime Minister, Zhao Zhiyang. The rooms were supplied with large vacuum flasks of hot water and a washstand in lieu of plumbing. The vacuum flask is of paramount importance in the domestic life of every Chinese and is used constantly everywhere, not least for the national obsession, tea.

Figure 8.2 Hostel life: the cook milling wheat flour for steamed bread (there is no tradition of baked bread) with a traditional quern right next to the coal heap and the bathhouse entrance.

Hot water was available on tap from two ancient coal-fired boilers; the larger, growling and rumbling ominously, also supplied the bathhouse. To get inside one had to negotiate the coal heap placed strategically outside the door. The wet floors reflected this hazard only too well and dropped soap entailed laboriously picking out bits of coal. Hot and cold emerged, however, but not always both together, from the bent pipes that served as showers. I never did get round to trying out the coffin-shaped concrete bathtub. The facilities were reserved for our exclusive use for 30 minutes each evening. Ablutions were accompanied by musical renditions from Ron and Alan who obviously thought China needed an introduction to Gilbert and Sullivan and rugby songs. The hostel pigs, whose sty was situated between the ladies and gents wings of the bathhouse, often joined in too.

The other essential facility was in a decorative little brick house in the grounds. The exterior was decidedly more prepossessing than the interior, which consisted of what might be described in cricketing terms as 'eight slips and a deep gully', with waist-high concrete dividers for modesty. Learning the pictograms for Ladies and Gents was an immediate priority. Typical Chinese lavatories are fine once you get used to balancing on two strategically positioned bricks. The 'night soil', as it is euphemistically called, is collected regularly and recycled onto the land throughout China.

Meals were taken in a small dining-hall. We ate in solitary splendour, separated from the Chinese by a row of screens. It was obvious that the cooks had gone to great lengths on our behalf. The diet was, of course, exclusively Sichuanese. Often we had little idea what we were eating, other than a suspicion that it was of vertebrate origin. It was wisest not to investigate too closely, bearing in mind the Chinese saying 'eat anything on four legs except the table, and anything that flies except an aeroplane'. The food was attractively garnished, good on the whole and for the most part well cooked, except for the case of the underdone sheep's testes, merely described as 'mutton,' so as not to offend our sensibilities, and the carp that came back to life under a coating of sweet-and-sour sauce.

The first day in Wangcang began, as did every subsequent one, with a rude awakening at 6.30 a.m. from the public-address system's earsplitting rendition of a repetitive military march 'Song of our country,' followed by an exercise routine counted to music. There was no fieldwork scheduled for day one so that we could recover from our travels. The whole party attended a formal welcoming ceremony in the mayor's office, perched on overstuffed sofas, consuming tea, tangerines and walnuts, and exchanging pleasantries with the mayor and county officials, in our case through Li Jinling. We learned that the town population was about 30 000, the whole county 800 000. The mayor was quite frank about overpopulation and unemployment problems. Indeed, people everywhere, dotted over the landscape, made a deep impression; no wonder that Chinese visitors to England remark on the emptiness of our countryside. The huge peasant population, organised into communes but divided into family groups, farmed the surrounding Dong He valley, a patchwork of tiny, intensively worked plots. There were also a few small coal mines and factories in the county. The town itself had a range of shops, a hospital, a cinema and television. It was even possible to telephone our nearest and dearest once a week from the postmaster's office, directly across the street from the hostel. It took anything up to several hours to make the necessary connections but the postmaster revelled in it. He had never telephoned abroad before.

Figure 8.3 Hostel life: vegetables, including cabbages and *bok choi*, were delivered daily and sold direct from the surrounding communes, prices being negotiated by weight (excluding the heavy carrying baskets).

Wangcang County held the inevitable banquet in our honour on the first evening. The officials were quite nervous, having never entertained foreign guests before. So far as anyone in the county knew, there was no record of any foreigners ever having visited it before we arrived. Thus we were the star attraction and a crowd gathered rapidly whenever we appeared in the town. Within seconds of entering shops, the post office or bank, we were surrounded by a crush of intensely curious locals who could hardly believe their eyes at the sight of three big-nosed,

Figure 8.4 Local industries were almost a step back to the Middle Ages. Ron Croucher wonders whether this lime kiln will produce suitable raw materials for our plaster-of-Paris field jackets.

strange-tongued creatures, head and shoulders above the throng. My pale eyes and fair hair also proved novel but, apart from our noses, Alan's beard caused most comment. Only very old Chinese grow facial hair, a thin wisp on the chin, so his hirsute visage was taken as symbolic of great age. He was accorded the special respect given to the aged, solicitously helped in and out of vehicles and had his field bag carried every day.

Two Wangcang County officials were attached to our party as 'logistics officers.' It soon became clear that their main function was to ensure our privacy and prevent the local people from becoming a nuisance in this respect. Nothing was ever said about leaving the hostel without an escort, they just made sure that none of us got through the gates unaccompanied. On one occasion, the three of us determined to escape undetected for a walk. Glancing furtively over our shoulders we sneaked out into the street, but our absence was soon discovered. Search parties were sent out and we were rounded up by Zhang the guard and a slightly reproachful Dong Zhiming. We did have a couple of escorted evenings out at the cinema, carefully surrounded on all sides by our party. The first was an ancient Chinese opera performed by the Hunan Opera. I could not even begin to follow the plot. Jinling, sitting next to me, confessed giggling that she couldn't either. The second was a total contrast, a Kung Fu epic *The Legend of Shao Lin Temple*. Filmed in Hong Kong with a mainland Chinese cast, including the Kung Fu (Wu Shu) champion, the story featured the chivalrous deeds of high-kicking monks set in the Tang Dynasty. Mercifully, it was subtitled in English.

The field locality was about 10 kilometres from the town and we commuted along the only, mostly metalled, road in an IVPP jeep driven by Lao (Old) Chen, not to be confused with Xiao (Young) Chen, who drove the lorry. Driving in China is not for the faint-hearted. The rule of the road is simple: whoever has possession of the middle wins. The jeep raced back and forth, honking furiously, scattering peacefully plodding water buffalo, pigs being walked on raffia leads and mule carts. Chen lent out of the window, constantly chastising pedestrians and cyclists, who were themselves suicidal, wandering and wobbling erratically from side to side across the road. On-coming vehicles approached each other at a frightening pace and swerved aside only at the last possible moment. Cars were a rarity, but clapped-out

Figure 8.5 Pedestrian hazards in the middle of the road were many and various.

buses, rickety tractors and smoke-belching lorries provided frequent heart-stopping moments.

Dinosaur bones and teeth have been weathering out their entombing rocks for millions of years. In ancient China they were thought to be the remains of dragons and were ground to powder for much-prized magic and medicinal potions. The first description of a dinosaur bone was written over 1700 years ago by a Chinese scholar, Chang Qu, who told of the discovery of dragon bones at Wucheng in what is now Sichuan Province. It is probable that Chinese dragon legends originate from the discovery of dinosaur and other vertebrate fossils, centuries before the concepts of evolution and the understanding that Earth's past history is contained in its rocks.

The fact that giant land-dwelling reptiles had once existed in the remote past was first realised by an English country doctor, Gideon Mantell, in the early 1820s. He noted that huge teeth from the local Cretaceous rocks closely resembled those of a modern iguana lizard and concluded that giant reptiles had existed in the Mesozoic era. Thus *Iguanodon* (iguana tooth), the most-abundant British dinosaur, got its name in 1825. The name dinosaur was coined a few years later, in 1842, by the famous Victorian anatomist and first director of NHM, Richard Owen, on the basis of three English dinosaurs, *Megalosaurus*, *Iguanodon* and *Hylaeosaurus.*

The first modern discovery of dinosaur remains in Sichuan dates back to 1915 when a large tooth and a thigh bone were found in late Jurassic rocks in Yung Si County by George Louderback, a Californian oil geologist prospecting for the then National Oil Administration of China. He took them back to the University of California at Berkeley where he was Professor of Geology. Twenty years later, in 1935, Charles Camp, Professor of Palaeontology at Berkeley, assigned them to a megalosaurid (a carnivorous theropod). Camp visited Sichuan in 1936 with Young Chung-Chien, the father of Chinese vertebrate palaeontology. Together they discovered Sichuan's first dinosaur skeleton. Since the 1930s, whole dinosaur faunas have been discovered in the Middle and Upper Jurassic beds of the Chongqing Red Basin. In those remote times 180–130 million years ago, the world looked very different. Central Asia (including China) was an isolated land mass surrounded by seas. The dinosaurs living there evolved separately, in parallel, with their cousins on the much larger supercontinent called Pangaea, which later broke up gradually to form the continents we know today. So, for example, the role of the whip-tailed diplodocid sauropods that inhabited North America, Europe and Africa was filled in China by the mamenchisaurs. *Mamenchisaurus* looks quite similar to *Diplodocus*, which stands resplendent in the NHM's Main Hall, greeting visitors with a jaunty flick of the tail. They both have small heads, long necks (*Mamenchisaurus* has the longest known neck of all) and long whip-like tails. But there are many detailed differences between them, which are clues to their separate paths of evolution.

Our site, on the northern rim of the Chongqing Red Basin, was situated in the picturesquely named Tangjia He (Soup Family River) valley, on a terraced hillside above a peasant farmhouse, and accessible only by a narrow footpath between rice and sweet potato fields. The farmhouse, inhabited by a garrulous Mrs Xu, served as base camp, storehouse and tea room. At first sight, the locality looked distinctly unpromising. There was bone there all right, eroding out all along the slope of the hill. But the bone-bearing layer dipped back into the hillside and was under anything from 0.5 to 2 metres of barren, hard grey-green sandstone (the overburden). Morale see-saws up and down on long field trips. Ours was rock bottom that evening when we three met for our life-saving coffee and postprandial nip of carefully hoarded duty-free. How could we possibly achieve anything useful here in just five weeks? Nine pairs of hands, no JCB, no bulldozer. I suppose we should have known better.

The IVPP team were quite used to tackling such large-scale operations. Dong Zhiming hired a gang of 15 peasants, including the local pig-doctor and carpenter, at 4 yuan per day – roughly £1, very good pay by Sichuan standards. The small wiry stature of the Sichuanese belies their strength and toughness. Within a few days they had shifted several tens of tons of rock from an area some 35 metres by 5 metres with nothing more than picks, shovels, small baskets – and dynamite.

The manual operation was aided by a portable electric 'Kango' hammer driven by a petrol generator that our side had contributed to the expedition. It proved ideal for a variety of tasks and a very popular tool. The gang all queued up to have a go, making light work of drilling charge holes ready for dynamiting the overburden. Explosives were readily obtainable, used frequently judging by the thuds that echoed daily around the hills, and not subject to any safety precautions whatsoever. I was somewhat disconcerted to discover what was inside those packages I had been using as a footrest as we bounced along in the jeep one morning! We stood gingerly by as Zhang Hong prepared the charges with a penknife, Her Majesty's Stationery Office masking tape and gay abandon, while Dong Zhiming squatted beside him, gesticulating impatiently with a handful of detonators.

Ron, having been a Health and Safety Officer in his time, was particularly horrified at the proceedings and expected to be blown out of his boots at any second. Hong had, of course, judged things just right; eight half-sticks at a time cracked and loosened the sandstone above the bones. Viewed from a safe distance, the most exciting moment was when one or two errant boulders landed on the farmhouse roof,

Figure 8.6 Our field site, a grey sandstone bed dipping into the hillside just above Mrs Xu's farmhouse in the Tangjiahe Valley. The excavated rock spilled over onto Mrs Xu's sweet potatoes for which she was generously compensated.

Figure 8.7 Standing at the top of the site, Li Jinling (left) and Zhang Hong (right) oversee the local gang removing the sandstone overburden loosened by the judicious use of dynamite.

shattering a few tiles and adding to the compensation bill already due for trampled sweet potatoes and general inconvenience.

We spent a profitable day or two prospecting for fossils up and down the valley while the gang did the hard work. Several promising future sites were located from which we collected isolated small dinosaur bones, crocodile teeth and large

Figure 8.8 Hard at work on site, members of the team, including Dong Zhiming brushing dust away from an articulated forelimb and Ron Croucher undercutting a plastered block, are watched over by Mr Zhang, the 'logistics officer'.

crimped crushing toothplates belonging to lungfishes. This proved quite a contrast to the lack of wildlife today in the area. I had little opportunity to recourse to my bird guide. It was easy to see why; gun-toting hunters with strings of small birds destined for the pot, combined with almost total agriculturalisation of all but the steepest crags and rocky hill tops, were obvious reasons for the scarcity of birds.

Our cleared site was quite spectacular, a bed dipping into the hill at some 25 degrees, packed with a jumbled mass of dinosaur bones ranging in size from massive limb bones 1.5 metres long down to teeth a mere 3 centimetres long. Most of the material consisted of completely dissociated skeletons with just the odd articulated forelimb here, a length of neck vertebrae there. Sauropod bones were the most abundant, followed by stegosaur and a few theropod elements. Bones were so closely packed that it was difficult to avoid treading on them. Inevitably, 15 pairs of peasant feet caused more than a little damage and frustration. The bone-bed extended over a far greater area than had been exposed by our operations and the local farmers were quite used to pieces of 'dragon bone' popping up in their fields. The deposit had formed from an accumulation of carcasses that had been transported down a river, washed up on a sandbank, decomposed and broken up. The remains may well have been scavenged too. The diners had left clues behind for posterity, because, among the bones, we found a scattering of dagger-like serrated-edged teeth, unmistakably those of carnivorous dinosaurs. Such 'mass-death assemblages' provide snapshots of dramatic natural disasters in the almost unimaginably remote past; but not so very different from the flood in same area just a year previously.

The work of mapping the site, exposing the bones fully with hammers and chisels, photographing, numbering and recording them on the map, jacketing and lifting, occupied the remainder of our time. The Chinese approach, based on the materials they had available, was somewhat different to ours. Some of the largest bones had crates built around them *in situ*. These were then filled with plaster and sawdust before the rock was cut away from beneath. This made them even heavier and carrying them down the steep hillside was both difficult and dangerous. The team of peasants managed these feats with the crates roped to parallel poles carried on their shoulders. Ron made their task slightly less onerous by wedging the crated bones with strategically positioned bags of plaster instead, to reduce the weight. He demonstrated and offered advice on a range of modern techniques and materials, using our supplies that had been shipped to Beijing in 1981. We experienced some difficulties with poor-quality local materials, particularly plaster that never set properly and reeked of hydrogen sulphide, and the low-grade solvents available for our glue and hardener.

We had managed all the while without the polyurethane foam. The Chinese railway authorities stood firm, even after a plea from the scientific attaché in Her Majesty's Embassy, Beijing. No, the chemicals could not be transported on a passenger train. Slow goods, the only alternative, would have taken anything up to two months with no guarantee of arrival at all. Luckily, the IVPP's deputy director, Sun Ailin, had planned to join us for the last few days of the excavation and, lo and behold, she arrived with two containers. A demonstration was finally managed on our penultimate day in the field. The technique involves mixing together two treacle-like components. These foam up and expand enormously to form a tough lightweight cocoon that looks rather like cinder toffee. The bone is first covered with kitchen foil to stop the foam sticking to it and a cardboard dam is placed around it. In goes the mixed polyurethane and five minutes later the jacket is done. Safety dictates that a respirator be worn while doing this because the foam gives off small quantities of cyanide gas as it cures. One of the Chinese preparators was so eager to try out a cocoon that he forgot to remove the paper seals on his respirator and nearly passed out through lack of air.

Figure 8.9 Lifting heavy crated specimens out of
the bottom of the site was a difficult task, even for
the tough Sichuanse with IVVP staff 'steering' at
the front and Dong Zhiming bringing up the rear.

Figure 8.10 Packing plastered bones in rice straw at Mrs Xu's farmhouse ready for road transport back to Beijing. The crates then had to be carried for several hundred metres down a narrow footpath to the single dirt road along the Tangjiahe Valley.

The weather was sunny and warm for the first week. Then it gradually got colder and wetter. At least that cured the mosquito problem in the little brick house but, on the other hand, my laundry took four days to dry. A total of six days work were lost because of impossibly muddy and slippery conditions on site. After one all-night thunderstorm and stair-rod rain, the bottom of the excavation was filled with water over a metre deep. Weather-bound days were long and boring, huddled in cold, damp rooms at the hostel. Jinling conducted Chinese lessons, patiently trying to accustom our ears to the unfamiliar four pronunciation tones. It is one thing to learn the right word but getting the tone wrong can lead to some serious social gaffes. It was so cold towards the end of our stay that the hostel eventually provided some heating. Placed in the middle of each room was a wooden stand supporting an iron plate into which was tipped a small heap of glowing charcoal. The net rise in temperature was minimal because the windows had to be opened to prevent carbon monoxide poisoning.

By the time we left, about 220 bones had been cocooned, lifted and packed with rice straw into 35 crates, all made to measure by the local carpenter. The IVPP team planned to stay another two weeks to clear the site completely. Our last day in the field was marked by a round of farewells to the peasant gang. Ron provided Polaroid portraits as souvenirs. Their reactions were a joy to behold as each was handed a blank square in which their faces gradually appeared.

After leaving Wangcang, we journeyed onward by rail, first to Chengdu, the capital of Sichuan Province, and then to Chongqing, to meet colleagues and to look at mid- and late-Jurassic dinosaur material, particularly in the Chongqing Natural History Museum. The next adventure was a three-day trip down the Chang Jiang (Yangtze River) by steamer from Chongqing to Wuhan, passing through the famous gorges. The last stage of our travels, by train inevitably, took us back to Hong Kong via Guangzhou, where we bade farewell to Li Jinling, who, brave and patient soul, had put up with us constantly for seven weeks. The first thing we did after arriving back in Hong Kong was to head straight for a steak house!

A complete replica skeleton of the late Jurassic Sichuanese stegosaur, *Tuojiangosaurus multispinus*, duly arrived at the NHM from Beijing. The only example outside China, it was mounted for permanent display in the museum's new dinosaur gallery in 1992. It was not the material that we had collected – that was far too fragmentary and remained in Beijing – but, nonetheless, a souvenir from the Orient of the most impressive kind.

Chapter nine

Close encounters in Pakistan

Peter J. Whybrow

Medical advice should be sought for all dog bites in countries where rabies exists as the treatment is urgent and somewhat complicated.

Preservation of Personal Health in Warm Climates, **1976**

In 1985, I was asked to assist the British Archaeological Mission to Pakistan. I was in a team of eight, organised by Robin Dennell and Helen Rendell in collaboration with the Geological Survey of Pakistan to search for fossils of human ancestors in an area close to the disputed border with Indian Kashmir. We had wandered into a deeply incised dry river bed and while clambering over house-size boulders found a beautifully preserved elephant skull that had not yet been rendered to silt by rare flash floods. Stony-hard it stared from the cobbles, too heavy to move but it gave a clue that near its grave might be rocks preserving more fossils from 2 million years ago when humans might have trundled around this part of Asia.

So where might these rocks be, higher up in the river bed or exposed in one of its several channels? The team agreed to undertake the 'seek-and-you-shall-find' method of palaeontological exploration and divided into scattered groups. I was a group of one and elected to climb the 3-metre-high river bank. Revealed was an amphitheatre with a backdrop of sheer cliffs that plunged to a broad, flat, grassy stage speckled with alpine-like spring flowers and heavily scented shrubs. The scene was wonderful; forget the fossils and savour the sun's warmth to lie in peace and harmony with Mother Earth in a Shangri La-like niche.

But others had found this place. Yap-yap, bark-bark rushed the dogs of war that guarded the gypsies' hidden encampment. The two sheepdog-like dogs were smart and I was their prey. To my front, one salivated with gross intent while the other circled to my rear. A lot has been written about what to do in such a situation – it's either staring out the animals while making placatory noises or throwing stuff at them. Staring out two dogs was difficult as my eyes do not function like a chameleon's and picking up stuff to throw would have brought me down to their level. So the inevitable happened. I used my only defence – a kick with a climbing boot. Standing on one leg is not a good defensive position. The canine war-school tacticians had obviously given tuition on dealing with bipeds and the dogs knew this ploy. A kick at one met merely with air while behind me the other bit at my lower leg. The wearing of baggy trousers helped (wearers of tight jeans take note). The bite was just with the front teeth but blood was drawn. As others of the team and the gypsies threw chunks of Pakistan at the beasts, I pondered whether the Reaper would call sooner rather than later.

Jhelum hospital at night is a sinister place. Feral dogs, more obsequious than those just encountered, slunk around its car park seeking food. What, I dared not to think about. Cha-cha, the Geological Survey of Pakistan's most-experienced driver (a septuagenarian, ex-British army), had taken me to Jhelum to have the first of a series of anti-rabies inoculations, a round trip of about three hours. We bumped along stony tracks, almost dry river beds and through agrarian villages, the populations of which had, like ourselves, never before seen such an antiquated Land Rover. Inside the hospital were long darkened corridors that had none of the frenetic activity of a British hospital. Having seen cars, trucks and buses on the Grand Trunk Road overtake three abreast I was puzzled. Were there no road accidents in this part of Pakistan; had all known disease in Asia been conquered; were the staff on strike?

In the Jhelum outpatients department, I was questioned by a matter-of-fact doctor who dramatically diminished my feel-good factor about Pakistan by telling me that rabies was endemic and that many had died from it.

'Was the dog rabid?'

'Don't know.'

'Did you catch the dog so that we can find out if it was rabid?'

'No.'

'Have you been inoculated against rabies?'

'No.'

'Well, you must have a course of injections lasting 14 days. These injections are in the muscle wall of the stomach. Report here every evening.'

'What, 14 days, in the stomach? You cannot be serious!' But he was.

On the seventh day I noticed several aspects of the hospital's facilities that led me to think that the Reaper was closer to me within the place rather than waiting at the end of the rabid road. At that time – during the rule of General Zia al Huq – Pakistan's main problem was a shortage of electricity, especially when it was most needed – at night. Power cuts were random but usually every evening for a few hours. During one session when a candle was the only light source, I saw that the vaccine was stored in a fridge that ceased to function during power cuts. Similarly, the syringe needle – seemingly the only one the hospital had – was sterilised each time in an

Figure 9.1 Jhelum.

electric boiler. Fearing hepatitis and other nasties more than rabies (the dogs were very fit and, at the time, not obviously rabid) I decided to end my outpatient visits to Jhelum and fully rejoin the team hunting for evidence of the earliest humans in Pakistan.

In the Potwar area of northern Pakistan during 1983, Robin and Helen had discovered stone tools fashioned from quartzite – a hard, quartz-rich sedimentary rock – firmly encased in a pebble sandstone, called a conglomerate. The problem was how to date the conglomerate. Mountains are born to be buried. Over the past 20 million years or so the eroding flank of the Himalayas has filled the northern part of the Indus River basin with 7000 metres of sandstones, clays and other rock debris. Such rocks do not easily provide material for modern dating techniques and the only tool available to the archaeologists was palaeomagnetic dating. At certain times during geological history, the Earth's magnetic poles have reversed. Sandstone can contain grains of an iron-based mineral and these grains preserve the fossilised imprint of the Earth's magnetic field at the time the sediments were being deposited. When patterns of reversal and normal events (normal is the Earth's present magnetic field) from samples taken from a specific site are plotted against the magnetic polarity time scale for the world as a whole, it is possible to get an estimate of the geological age for the samples.

Rock samples collected from above and below the conglomerate containing the artefacts produced a palaeomagnetic 'date' of between 700 000 and 2.48 million years – not very precise. Consequently, information of the time when the rocks in the area were folded was used in conjunction with the palaeomagnetic data. From this analysis the British Archaeology Mission scientists concluded that the minimum age of the rocks encasing the artefacts was 2 million years.

This was a mind-blowing conclusion. The London *Times* of 1987 reported: '2-million-year-old axes change human dating. The discovery of primitive axes in a remote corner of Pakistan has prompted scientists to call into question the whole chronology of the evolution and dispersal of early humans both in Asia and Africa.'

The publicity was not well received by academe. Articles by opponents of the idea that humans were present in Asia at the same time as in Africa stated, 'in the absence of evidence of hominoid presence in Asia so early, the claim that they are artefacts must be doubted;' also, 'Serious problems regarding the identification of the man-made objects are common in Lower Palaeolithic research and often underestimated;' and, 'None of these criteria [those used to identify the quartzitic rock as a human artefact] is sufficient to permit reliable identification . . ., especially for the stratigraphic context of these finds – a coarse conglomerate, most probably fluviatile in origin.' The British Archaeological Mission team replied to these criticisms in detail but there remained the puzzle that if hominoids had made the 'artefacts,' where were their fossilised remains?

Expeditions to the Pabbi Hills, about 140 kilometres south-west of Rawlpindi, started in 1987. Near the delightful village of Kharian, a place where noise from the Grand Trunk Road mixed with smells and insects from the open sewers and scents from bakers, tea shops, butchers and hairdressers, the team's camp was a deserted, dilapidated bungalow, Bannu Bangla. Probably about 100 years old, it had been built near an ancient Mughul reservoir. About an hour's walk from it, Robin had found a place under some enormous thorn bushes where a substantial amount of fossilised bone had weathered out. This pile, it was believed, represented a 1-million-year-old hyaena burrow and the remains of ancient humans might have been incorporated into its spoil.

A detailed excavation was necessary. Every day for two weeks we walked to the site by clambering down cliffs, dodging thorn bushes, toiling up watersheds and weaving down the tortuous path formed by dry stream beds. The easiest part of the walk was along a flat, steep-sided, tree-lined track – rather like a country lane in Devon. But in Devon, short, squat, vicious-looking mongrel Pit Bull terriers – without ears – do not stand in the centre of the lane daring anyone to pass them by. In Pakistan, these dogs are used for fighting, hence their lack of ears. We skirted this area for several days, becoming cut and punctured by thorns in the process and, mindful of my 'attack' in 1985, loaded our pockets with defensive stones.

Excavation was hot, dirty and difficult as the bones were exposed on a steep, three-sided plateau no larger than a billiard table. The long walk back to Bannu Bangla each evening after our contorted day with trowel, pick, hammer and surveying equipment became a chore, and the team became tired and irritable. Lack of funds to purchase adequate supplies of imported Chinese toilet paper added to irritations (not that there were any toilets) and most of the team experienced thorns in places that normally thorns should not reach. In addition, nightly but brief, torrential rains seeped through the buffalo-dung clad roof of our bungalow to drip – like a Chinese torture – upon us. Then, Hannah Jones told us she had discovered a shower in Kharian, that just happened to be at the rear of a barber shop. Some of us were sceptical, but we needn't have worried. Our showers became the high point of the trip and everyone worked harder so as to have more time in its luxury. Also, our visits to the village became the high point of anything that had happened in Kharian since 1947, apart, that is, from the Kashmir dispute.

Entrance to the shower facility was through a badly draped curtain after passing through the hair-cutting salon where two chaps attended the tonsorial grooming of Kharian's men. The curtain separated this area from six concrete cubicles each with a crude toilet-like door. Hot (sometimes) and cold water pipes with huge valves bent and swayed over the cubicles and a courtesy bar of abrasive Chinese soap rested on the top of the partitions, which were more than 2 metres high – difficult, therefore, for some team members to reach.

At first, our visits caused little interest. Then the Call must have been broadcast, 'The Europeans are about to shower.' Five of our team were female. The barber shop was never so busy as when team members armed with extra towels, shampoo and other impedimenta trooped through the male throng. Males of the team waited patiently but had to engage in long conversations, continuing the next evening, about jobs in the UK: Did we know someone's father in Bradford? Why was Mrs Thatcher charging Pakistanis for an entry visa? Do you know that the headman of Kharian used to drive a bus in High Wycombe? And do you want to buy an AK47? Meanwhile, the first in the queue, Hannah, whose demure Scandinavian looks had

Figure 9.2 Excavation of the 'hyaena den'
at Kharian.

previously caused one astonished butcher in Kharian's fly-ridden meat market to chop off his finger tip, was enthusiastically greeted. She herself, while reaching for the soap, noticed that a mirror artfully placed at a great height in the barbers gave both barbers and customers a limited but tantalising view of the team's ablutions. Some of us wondered if we could supplement the expedition's meagre resources by charging to view The Mirror.

Since those few delightful days in the Kharian region, I have lost contact with the team, most of whom were active graduate students. No astounding finds of fossilised humans have since been reported from Pakistan. Money for such projects tends to diminish when objectives are not achieved. On the other hand, political constraints in Pakistan can interfere with other initiatives.

Sulaiman Mountains 1989

Kashmirwalas Hotel faces Rawalpindi Mall, 'Pindi's cricket pitch, and is the place where tourists stay who cannot afford the luxury of Flashman's Hotel or the even more expensive and pretentious Holiday Inn. This was my first stay in Kashmirwalas. The Silver Grill, a minimalist but a wonderfully hospitable hotel that many palaeontologists had used as a first base on arrival in Pakistan, had been sold and demolished to make way for a multistorey, marble-clad building to house the now defunct and notorious Bank of Commerce and Credit International – BCCI.

I met with Mahmood Raza, Assistant Director of the Geological Survey of Pakistan (GSP), who had just returned to Rawalpindi after carrying out collaborative studies on the vertebrate fossils found in the Salt Range with the Harvard University group led by John Barry. John and his team had been working the area since 1973 and their study is a palaeontological monument to the painstaking and detailed work necessary in obtaining evidence of the past life of Pakistan's fossil vertebrate fauna from some 20 million years ago. Mahmood arrived with Phil Hurst, BP Exploration, and Peter Friend, University of Cambridge, both of whom had been studying the morphologies of the great sandstones, more than 1 kilometre thick, in the Salt Range, Phil to get an idea of their sedimentary geometry for the purposes of finding hydrocarbons and Peter wanting an insight on how, where and when the ancient Indus deposited such a vast thickness of sediments.

My objective was to seek assistance for a reconnaissance trip to the eastern flank of the Sulaiman Range some 450 kilometres south-west of Rawalpindi. Here, since the nineteenth century, Miocene outcrops had produced an interesting fossil fauna of large mammals, rodents, reptiles and fish, some of whose ancestors might be linked with similar animals I had previously collected in Saudi Arabia. The idea, if the work could be carried out, was to obtain more evidence of certain mammals that might have dispersed between Arabia and south-western Asia about 20 million years ago.

After the usual bureaucratic delays, this time compounded by the Director General of GSP who wanted Mahmood to justify his involvement with foreign teams, I travelled to Dera Ghazi Khan with Khalid Shah, GSP's budding geochronologist. Passing through the Indus Valley I saw how rich the agricultural landlords were and how poor the government must be. Roads, electricity supply and other items of Pakistan's infrastructure were decaying, yet the profit from agriculture must be high judging from the many wealthy residences occupying the suburbs of Dera Ghazi Khan, that, oddly, boasted a Chinese restaurant. That evening Khalid took me to visit an old friend of his from Peshawar University. We entered his house through a tiny gate in the high wall of one of the up-market residences and were ushered into a massive lounge equipped with TV, video, music centres, other modern electrical nick-nacks, air-conditioning and expensive modern European furniture. Khalid and his friend chatted away in Urdu and eventually I managed a question: 'Do you own a farm?' 'No,' was the answer, 'I work for Pakistan's Atomic Energy Commission.' Many aspects of life in Dera Ghazi Khan suddenly became clearer to me.

The area I wanted to visit was near a village called Vihowa, close to the border of Punjab, Baluchistan and the southern part of the North West Frontier Province. Vihowa itself lay on the Indus Plain, just east of a pass through the Sulaiman Mountains to Afghanistan. This area was called 'difficult' as the route was used by drug runners and other smugglers from the Golden Crescent. Consequently, we had to have an escort and we met with a young man from the Baluchi Levy. I asked what the rifle was for and he, not unreasonably, said it was for shooting people. When asked if the one ammunition clip was enough to shoot most people, he didn't answer. The next morning it was chilly and I was wearing a jacket comprising mainly of pockets for such things as passport, money, hand lens, pocket knife and useless objects thought essential to bring from the UK. We picked up our minder from his house in our tiny Toyota jeep. He presented me with at least ten ammunition clips to stuff in my pockets, 'To shoot many people with,' he laughed.

Suitably protected from hoards of drug traffickers, we left Dera Ghazi Khan bouncing our way to Vihowa across numerous salt-encrusted fields, the result of overirrigation to produce short-term cash crops and, later, attempted to overtake buses and tractors detouring around what was once the main road, breached every 800

Figure 9.3 Members of the Baluchi Levy.

metres or so by flood waters from the mountains. Extremely sore from the ammunition clips that had been poking into my ribs and thighs, we arrived at Vihowa's Victorian police post. The Subbadar, Sayed Amir Mohammed Shah, and the Naib Subbadar, Ashiq Hussain Kulachi, both of the Baluchi Levy, sent out for a delicious chicken biriani, but before it arrived, an embarrassingly difficult two-hour long stunted conversation loaded with local politics took place. The Baluchis thought themselves a cut-above Khalid from GSP (a Punjabi), who, in turn, thought himself better than the mountain men. All I had was a smattering of Arabic.

Khalid lazily translated the Subbadar's questions from behind Pakistan's equivalent of *The Times*. 'Are you from an oil company ?' 'No, I hope to collaborate with GSP to search for fossils in the hills around here.' 'Ah, we shoot you British when you come with horses into pass. You come on and on in straight line and we just run behind rocks and shoot again.' Pleased that I did not come to Vihowa on horse I asked Khalid if I should keep the ammunition in case our minder wanted to kill us both. 'No,' he responded, 'the Subbadar is actually telling you about his great-grandfather's exploits with you British during the time of one of your punitive raids into Baluchistan.' Enquiring, via Khalid, 'Who won?,' the Subbadar exploded into laughter, thumped

Figure 9.4 Negotiating a boulder strewn river bank formed by flash-floods.

me on the back and, apparently, said 'Oh, no one – it was a draw – just like playing cricket.'

The next day we by-passed the main river up a steep track strewn with ever-increasing numbers of cobbles and boulders. Walking was faster, and less back-breaking, than the speed of our toy-like transport that perambulated the track. We moved larger boulders out of its path and eventually breasted a hill to find the main river bed. The force and volume of mountain water over thousands of years coupled with the uplift of the mountains themselves had created a massive gorge along which we then sped. Passing by exposures of increasingly older rocks, we headed deeper into the mountains until we reached an area where the gorge narrowed and only walkers could go forward. Climbing up cliffs we soon found fossilised molluscs, black-coloured mammal, crocodile and turtle bones, while our policeman found a valley strewn with large chunks of silicified trees. The whole of this region comprised a series of parallel ridges, separated by steep valleys, that marched back through geological time to the centre of the Sulaiman range – in geological terms, the flank of an anticline. It seemed an excellent area to plan a thorough reconnaissance for a long-term project. I discussed this idea with the Subbadar at the forward police post – he had come later, on horse, of course – and he immediately proffered all forms of hospitality for my continued visits, together with adequate protection from bandits wanting to steal our fossils.

Figure 9.5 The wadi-bed road to the mountains.

But it was not to be. I had obtained a grant from an oil company for the project and had discussed this with Mahmood whose GSP team would benefit through the hire of equipment and facilities. His problem was that other palaeontological teams, mainly American, had also requested help to work in the Sulaiman Mountains and he could foresee GSP's limited resources being thinly stretched. On the other hand, he wanted to work with us and the issue became one of prioritising the various requests for GSP involvement – perhaps whose team would most benefit GSP in terms of dollars? The other teams got to hear of this and their letters to Mahmood the following year caused him to reassess his involvement. This coincided with the political problems in Pakistan experienced by many international organisations, such as oil companies, under the new government of Benazir Bhutto. Eventually an

Figure 9.6 The policeman from the Baluchi Levy
with his find of silicified wood.

agreement enabled my grant allocated for Pakistan to be transferred to work in the Republic of Yemen, where, following the 'marriage' of north and south, it was hoped that southern Yemen would become as rich in oil as Saudi Arabia.

Ironically, the area around Vihowa today is reported to be buzzing with oil-drilling crews, seismologists, other professionals and equipment associated with a major hydrocarbon exploration programme. Perhaps the Subbadar from the Baluchi Levy is still telling stories of his ancestors' exploits to the geologists from British Gas?

Figure 9.7 Flank of an anticline in the
                Sulaiman Range.

Chapter ten

The day of a thousand fossils

Peter Andrews

In no sphere of archaeological or anthropological research are such startling discoveries being made as in the Ancient East.

The Most Ancient East,
V. Gordon Childe, 1928

There are two ways out of Africa. One way is to go eastwards out of Somalia or Ethiopia, across the Arabian desert and into Iran and points eastward. The other way is to continue northwards through Egypt, Israel, Palestine and thence to Turkey. From Turkey the whole of the Old World is on your doorstep, east, west or further north. That is the reason why I decided, on leaving my job in Kenya and taking up a new appointment at The Natural History Museum (NHM) in London, that I would look for fossils in Turkey.

I had been working for almost ten years in East Africa, trying to find fossils of hominoids. The hominoid group comprises all extinct and living apes and humans. Therefore hominoid fossils can shed light on our own ancestry as well as that of chimpanzees, gorillas and other apes. My work in Africa involved describing a large group of fossil apes from Kenya and Uganda, in collaboration with Dr Louis Leakey until his death in 1972. There is no doubt that apes evolved in Africa before 20 million years ago, but what was not so clear was what happened to them after they left Africa. This is what I wanted to find out by excavating in Turkey. I thought that by looking there for fossils in a particular time band I would find apes at a stage just after they had left Africa. Then, by comparing these fossils with those known from Europe and Asia, I would be able to understand more about ape evolution. That was my idea. But, when I started work, only one fragmentary ape fossil was known from Turkey, so my chances of success seemed slim.

Fossil apes had been known from Europe and Asia for some time. One type of fossil ape, called *Dryopithecus*, from the south of France was described as long ago as 1856, three years before Darwin published his *Origin of Species*, and in the early 1900s other fossil apes were found in India. As other fossils were found, it became clear that apes were once widespread across Europe and Asia, but it was not known how many species there were, what they looked like, how they lived nor how they were related to living apes. Turkey seemed the obvious place to go to try to find some answers.

I first went to Turkey in 1976 with a group of colleagues from the NHM. We excavated a couple of sites in the south-west part of the country, but although we found many fossil mammals and sorted out many aspects of the local geology, disappointingly we did not find any fossil apes. At the end of the season, I did not return to England with my museum colleagues, but instead went on an exploratory trip with my wife Libby. We hired a car and drove up the Aegean coast of Turkey through Izmir and Balikesir looking for a tiny village in the hills called Pasalar. It was not marked on any of my maps, but a German friend who had worked there previously had given me rough directions about where it was. More by good luck than good management we eventually found the village, and there our troubles began, for there was no road to the site. We drove our car up and down the cart track that was the main road of the village. At last we decided that the fossil site was likely to be down in the valley rather than up the slopes of the high mountains surrounding three sides of the village, so we set off through gardens and along goat tracks until we came to the river, where we immediately got stuck. Luckily there were lots of people around who were remarkably patient and polite to the foreign intruders, and they pushed us out of the water so that we could continue our trail. Further along another goat track we found ourselves in a most beautiful valley driving along a new road still being made, with forests of oak and beech on either side of a lovely mountain stream. There was an old water mill at a bend in the road, with a mill race at one side, and just there, in a road cut on one side, was the fossil deposit.

It was extraordinary. Never have I seen so many fossils just falling out of the side of a hill. In only one hour we found specimens from more than 20 different mammalian species, but no hominoids. Some were found in great abundance, mostly represented by isolated teeth but there were many bones as well. We did not have a permit to collect, but we made a discreet pile of fossils and carefully placed them so that when we returned next year, as we optimistically hoped (with proper research clearance), we could immediately pick them up. After this we had to hurry back as it was getting towards evening and we had a long way to go to reach the nearest town. Again, our ill-judged optimism let us down; we got lost on a road that led interminably up one side of a mountain, where there should have been no mountain at all, and then down the other without ever seeming to go anywhere. To this day I still do not know where we went as there were no sign posts to say what the place was called.

Figure 10.1 Looking up the Pasalar valley from Pasalar village.

The next year, 1977, came and went, and although we made all the preparations for working at Pasalar, including buying air tickets from London to Istanbul, we did not get our permits. It was not until 1983, six years later, that I finally managed to return to Pasalar with my colleague and good friend, Berna Alpagut, Professor of Anthropology at the University of Ankara. We started with a small team, I with just one student and Berna with a colleague and good friend, Erksin Ersoy. That first year we prospected the site, drew plans and dug a single trench through the deposit, not looking particularly hard for fossils but finding no less than 23 hominoid teeth in the trench and on the surface.

I returned to Pasalar in 1984, this year to conduct a proper excavation. My student Lawrence Martin had now completed his doctoral thesis and came again as a full member of the team. I had worked with Lawrence on several digs before, the most notable one being an excavation in western Kenya. Our English colleague in Nairobi, with whom we were supposed to be collaborating, had found a fossil-rich deposit at a site called Meswa Bridge. After starting an excavation at the end of the field season Lawrence and I applied for funding to return there the following year to complete the excavation. I sent Lawrence out early to make all the necessary preparations for excavation, but, for reasons we could not understand at first, he met with a certain resistance from our 'colleague.' This was disturbing to say the least, because we had

Figure 10.2 Myself and Berna Alpagut in the first
year's trench excavating an elephant tusk.

received a big grant to excavate the site. Worse was to come. We arrived at the site to find it all gone except for a few remnants around the side of an empty pit. The emotions I experienced standing by the side of that devastation have remained with me all my life. What had happened was that our Nairobi 'colleague' could not wait for our arrival but had hired a gang of workmen to pickaxe out the site and had carried several massive blocks of sediment to Nairobi. Naturally his enthusiasm for our collaboration had diminished. Matters were quite tense for a while, but we eventually managed to locate sufficient remaining pockets of fossil-bearing sediment to make our expedition worthwhile, although this remains one of the great 'might have beens' in my anthropological career. We were joined at Pasalar by two promising young students from the University of Ankara, Ayhan Ersoy and Songul Alpaslan, both now with completed PhDs. We started our excavation at the south end of the deposits, which turned out to be a bad choice because this proved to be an area of intense spring activity. The fossil-bearing deposits rest on a massive bed of limestone, through which water flows up from pressure higher up the mountains and comes to the surface as a series of springs – in fact, one of the springs still active today supplies our drinking water. These springs used to flow out through the fossiliferous deposits, dissolving away the bone; hence complete tooth rows were left in place but without the jaw bones in which the teeth were positioned.

I continued work at Pasalar the following year and every year to the end of the 1980s. Lawrence continued with me, sometimes with his wife Wendy, and he and I used to bore other people at the camp because we talked endlessly about fossil apes. Lawrence did his doctoral thesis on the Pasalar apes, looking at the thickness and structure of the enamel covering of the teeth, a pioneering study that has still to be matched from any other fossil site. Lawrence and I worked together for many years, although now he has moved more into university administration where he is doing very well.

Our excavation at Pasalar followed normal palaeontological procedures, with a metre grid laid out over the whole site tied in to a single datum point. Each grid square was individually excavated, with one person digging in each square usually for the whole of one field season. He or she would dig with a small hand trowel or pick, brushing up the loosened sediment into buckets to be taken across the road and sieved in running water. The fossils were mainly in the sandy layer, and, to reach this, the overlying soil and sediment had to be removed, sometimes laboriously by hand and sometimes with the help of a mechanical digger, which provided plenty of excitement when the driver got carried away and tried his hand at excavating fossils with his enormous shovel.

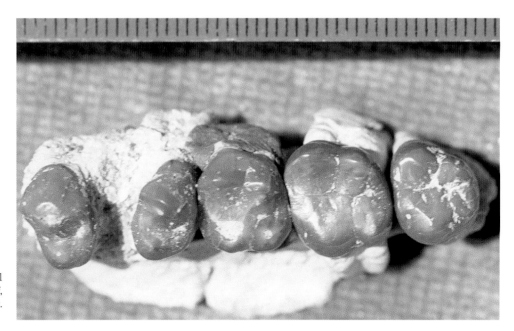

Figure 10.3 One of the upper jaws of the fossil ape, which we later called *Griphopithecus alpani*, from Pasalar.

In 1986, we began to find more-complete jaw specimens of a fossil hominoid, which at that time we called *Sivapithecus darwini*. Both the teeth and the jaws resembled the fossil ape found in India in the 1900s and named *Sivapithecus*. We thought our Turkish ape was related to the Indian one. The Indian ape, as I had just shown in a paper published in *Nature*, was related to the orang-utan, the only living great ape from Borneo and Sumatra in eastern Asia. The geological date of Pasalar is important here, for we had concluded that the site was about 13–14 million years old, which made the ape from Turkey the earliest from Eurasia known at that time and put it at least 1–2 million years earlier than *Sivapithecus* in India and Pakistan.

What we were proposing was that the fossil ape from Turkey was ancestral to the sivapithecine and orang-utan lineage in Asia and signalled its origin. This was quite a claim, for at that time (and it remains true to this day) we had no fossil evidence for the origin of any hominoid lineage other than that leading to human beings. I still think that the relationship we proposed is correct, although now, with more material to hand, Lawrence and I have separated the Pasalar ape from the Asian ape *Sivapithecus* and we now call it *Griphopithecus alpani*.

One of the first essentials to seeing the fossiliferous deposits at Pasalar in a wider context was to look at the surrounding country. In particular I wanted to find out if there were more deposits like the ones we were working in. I also wanted to discover the source of the sediments, for our fossil-rich sediments were quite different from the other sediments in the valley. Berna and I spent many happy hours walking and riding

through some of the loveliest country in Turkey, along mountain ridges, up mountain streams and waterfalls, and into hidden valleys. We had managed to borrow a top secret map of the area, which we had been told to guard with our lives as it provided much more detailed information than the standard maps. One day we had just got to the top of one of the highest mountains and were feeling quite pleased with ourselves, when we were brought down to Earth very quickly, for on placing the map on a flat rock to see it better it was suddenly blown away over a cliff and into a valley at least 400 metres below. We never found any trace of it, and, needless to say, were never lent another.

Our investigations were not without hazards, some of them dangerous. We got lost more than once and missed more than one meal as we struggled to find our way out of the maze of valleys and hills. This area of Turkey is unfortunately being developed at present. One day, as we were driving along the lower slopes of a mountain called Kiziltepe on a road that looked half-made, we found that half was the right word, for as we came round a corner we heard a terrific explosion and rocks the size of our car suddenly started falling right in front of us. The road workers were still blasting out a course for the road, and we had come unexpectedly on the scene of their operations. Our driver put the car into reverse and went backwards around a hairpin bend without even looking, so that I was never sure if we were in more danger of being crushed by falling rocks or being driven over the unprotected side of the mountain into the gorge hundreds of feet below.

These explorations bore fruit. We found that most of the mountains in the area around the fossil site were made up of igneous rock, and this was the source for all the higher sediments in the Pasalar valley, but it was totally absent from the fossiliferous deposits. These were composed of pieces of marble and other pieces of metamorphic rock that today occur along narrow ridges making up the sides of the valley immediately above the fossil deposits. These sediments accumulated, therefore, from the decay and weathering of these rocks close to the site, so that I could be confident that the sediments were local in origin rather than washed in from great distances. As we shall see, this discovery helped greatly in interpreting the history of the site.

As the field work at Pasalar progressed, Berna and I were joined by various specialist workers. A palaeontologist from Finland, Mikael Fortelius, was one of the first and worked with us for several years. He was a good colleague and did a marvellous job identifying fossil mammals and describing groups like rhinos and pigs. The first essential in working in a country like Turkey, which has a highly sophisticated culture very different from that of western Europe and North America,

Figure 10.4 (*Opposite*) The mechanical digger at work at Pasalar, removing overburden and digging a trench (left) through the deposits.

Figure 10.5 The Pasalar deposits, showing
part of a rhino jaw being excavated on the surface
and numerous bones exposed in section in
the foreground.

is to accommodate one's behaviour to local custom. I was not always able to do this successfully, which led to several spectacular rows with Berna, but she and I were always able to talk things through eventually and reach an understanding about problems in communication and differences in behaviour.

Erick Bestland, an American geologist, spent one season with us at Pasalar, and his detailed analysis of the physical and chemical structure of the sediments helped enormously with my understanding of the site. Another American, Douglas Ekart showed his liking for Turkish people, especially the female students, in the clearest possible ways. He was a brilliant student, who also worked with me at another Miocene site in Hungary, but he hid his abilities in the most effective way by never writing anything about his work. The first time he came to Pasalar, he did not let anyone know he was coming and, in fact, arrived several days before anyone else. Somehow without knowing a word of Turkish (and no one in the village knew any English) he managed to make friends with Refik, the mill owner and a long-term supporter of our work, who put him up in his house in the village and kept him entertained, or was it the other way round, until the excavation team arrived.

Refik was a lovely man, now sadly dead at a relatively young age. He owned the mill house and operated the water mill for the local farmers in the region. His wife made us a loaf of bread once from the stone-ground flour produced by the huge grinding stone in the mill, and the plentiful supply of ground-up stone mixed in with the flour laid the whole camp low for a couple of days. In fact, it was only the foreigners who were affected, for the Turkish members of the team were wise enough not to attempt to eat the loaf, preferring the marvellous locally produced white bread. The dig has taken over the whole of the mill house now that it is no longer operating, so we have the luxury of a massive living, dining and cooking room with water laid on (from the mill race), electricity brought up from the village 2 kilometres away and the telephone (a mixed blessing).

Over the years we have continued to find more and more hominoid specimens, as many as 160 in 1992. Bones from all sizes of mammals of many different species are preserved at Pasalar. Hominoids, unusually, are among the commonest fossils – at most fossil sites of similar age they tend to be much rarer. We found skulls and lower jaw bones of two species of elephant and other remains of several species of rhino, giraffes and antelopes, many carnivores such as bears and cats and many other medium-sized and small mammals. A small gazelle-like antelope, represented by many jaws and partial skulls with horn cores, is the commonest species; it browsed on the leaves of bushes and trees growing in the area. A large type of extinct pig with tapir-like

Figure 10.6 Refik standing in front of the mill house with his wife, who is holding the loaf of bread.

Figure 10.7 (*Opposite*) The mill house in the foreground with the fossil-bearing deposits at Pasalar just visible as the white sediments in the middle background.

teeth is also common. Next commonest are the hominoid primates, either one or two species – we are not quite sure. The remains of the hominoids, mostly teeth though we have found a few fragmentary limb bones and jaw bones, are more broken up than those of the antelopes and pigs of similar size. Fossils of many rodents and lagomorphs (akin to rabbits) are also common at Pasalar, and we found a type of extinct beaver as well.

So far we have recorded 58 mammalian species at Pasalar. None of them is still around today so that details of their lifestyles are unknown. Most of them are related to living animals, however, so that we can make some inferences about the environment when the fossil-bearing deposits were laid down. For example, the presence of tragulids (small browsing hoofed mammals or ungulates related to living chevrotains) might indicate similar habitats for the fossil species – namely wet, swampy tropical forest.

Another way of looking at the environmental issue is to ask what adaptations the fossil animals had. Like the tragulids, many of the plant-eating mammals, including the ungulates, were browsers, and even the single species of horse had low-crowned teeth that were unsuitable for grazing (modern horses have high-

Figure 10.8 Part of an elephant skull being excavated.

crowned teeth). One species of antelope, one of the rhinos and the two species of lagomorph may have been half-grazers and half-browsers. The preponderance of browsing species suggests that the Pasalar locality was densely wooded rather than open grassland. Most of the rodents and the primates were adapted for a fruit diet, while the pigs and some of the carnivores, such as primitive bears, seem to have been adapted for eating a variety of food; they were truly omnivorous (like we are). This spectrum of feeding habits is typical of mammalian communities in tropical forests today, where a variety of food is available on the ground and in the trees.

If we now look at how the Pasalar mammals lived, we find that there were few wholly tree-living animals; the greatest number partly lived on the ground and partly scrambled about in trees. This lifestyle and the size range of the animals suggests that the Pasalar area was subtropical forest in middle Miocene times, around 13 million years ago, an environment characterised today by highly seasonal summer rainfall, such as one gets in the monsoon forests of India. And the vegetation probably consisted of seasonal forest, probably fully deciduous.

One way of trying to find out more about the environment at Pasalar 13 million years ago is to look at modern habitats to see what kinds of animal they have in them and to compare this information with what we know of the Pasalar fauna. I went on one trip to a forest in western Turkey with a hospitable local business man called Ismail Bey, who was always more than ready to help with trips like this. In this particular case he brought along a few 'optional extras' such as a full-scale picnic and a five-man band, and we spent all afternoon dancing to the band and sampling the food and drink, which was in plentiful supply. By the end of the afternoon I was convinced that this forest was just what the Miocene apes would have most appreciated.

At the end of the 1989 season, I decided it was time to publish the first results from our research at Pasalar. This covered the first seven-years work there, and I invited additional specialists to describe some of the animal groups such as the rodents and other small mammals. There is a sad story involved here. We spent the whole of one season processing large quantities of sediment from all parts of the site, using graded nests of sieves to collect a good sample of small mammals such as rats, mice and shrews. In this way we hoped to record their distribution at the site. Part of this collection was shipped to colleagues in America for sorting and identification but the remainder of the material was left in Turkey, where a student was delegated the job of processing it. The student was instructed to replace the field bags with stronger ones for long-term storage. Unfortunately this instruction was misunderstood, and all the samples from all over the site were put together into several large storage sacks, thus thoroughly mixing up the samples that we had been at such pains to keep separate. Goodbye small mammals!

The results of our work at Pasalar were published in 1990. I managed to get the agreement of the editor of the *Journal of Human Evolution* to publish our results as a special issue of the journal. This was not difficult because the editor was myself, but I gave myself a hard time by insisting on a full coverage of the site. Most chapters were on the palaeontology of the different animal groups, but there were also chapters on the geology, taphonomy and palaeoecology. Taphonomy is the study of how bones are preserved as fossils, and this formed an important part of the work at Pasalar, partly because the site is so complicated and partly because I am also a taphonomist. I will outline some of our taphonomic evidence before finishing with a short reconstruction of what I think the Pasalar area was like 13 millions of years ago.

One feature of our fossil collection from Pasalar is the above-average number of immature animals in the assemblage, especially of large species such as rhinos and giraffes. Today, few large cats and other carnivores can kill fully grown rhinos or giraffes (let alone elephants), but they will tackle the young of these species and carry off parts of the carcasses to their dens. This is what I think happened at Pasalar. The same carnivore species also prey on adults of smaller mammals, so that mature bones from these species are mixed with the immature remains of the large species. Again, this is exactly the situation at Pasalar.

The action of predators on the way the bones of dead animals are dispersed is the first stage of 'taphonomic modification.' Later modifications arise after bones are exposed on the surface of the ground. A few bones are buried quickly, perhaps by a burrowing animal inadvertently covering them up with soil, so that they stay fresh and unbroken, but most remain exposed for some time. There they are broken and weathered by exposure to the sun and rain and by trampling by large animals. Looking at how bones weather today, I think that most of the Pasalar bones remained on the surface for periods of up to 20–30 years before being buried. Even before they were exposed some had been broken up by predators – hyaenas, for example, crunch up limb bones to get at the juicy marrow inside. Not surprisingly, therefore, few complete bones fossilise, and this is what we found at Pasalar.

I think that the Pasalar bones were then caught up in a massive mud slide and washed down off the hill slopes and dumped on to what is now our fossil site but was then a swampy area of silts watered by springs. The springs welled up through the limestone at Pasalar, creating large depressions in the mantling silty sediments and softening the surface of the silts, so that trampling by large mammals, attracted there by the abundant water, made deep potholes into the surface. As the mud flow was dumped on to this uneven surface, larger rocks were caught up by the potholes and some parts of the underlying silts were dragged up into the mud-flow deposits. While all this went on, the springs continued to flow, producing inverted cone-shaped deposits as the springs were filled with the mud flow.

During the movement and deposition of the mud flow, the larger bones and stones became concentrated towards the bottom of the deposits and the smaller ones nearer the top. Thus, at Pasalar today, we found most of the large mammals near the bottom, often indeed in the potholes themselves, and rodents and other small mammals near the top of the excavation.

What events produced such an unusual fossil site as the one at Pasalar? Its unique features include the huge numbers of fossils concentrated in a small place, no more than 20 metres by 14 metres, the differences between the fossiliferous sediments and the overlying sediments and the many features of predation, weathering and transport that were so important in bringing the bones together.

What follows is my story of the day of the thousand (actually hundreds of thousands) fossils.

About 14 million years ago, it was a hot and humid day on the slopes of the densely wooded mountain in western Turkey. The temperature was well over 40 °C. For the previous few days the clouds had been building up and, with them, the humidity until every movement was an effort. The animals in the valley and on the hillsides moved restlessly, and those that could started moving up the hill slopes seeking cooler air above. In the late afternoon the rains began, sweeping down over hills and valley and driving the animals still higher up the slopes away from the threatening floods. It rained incessantly, soaking into the thin soil on the hills and flooding the swampy valley bottom. Finally the rain penetrated through to the limestone rock under the thin hillside soil. The surface of the limestone, long since metamorphosed into deeply folded marble, parallelled the hill slope along the south-facing side of the valley, and when water began to run over the rock, underneath the soil, it took the soil with it down the slope. Gradually, with increasing momentum, the slippage turned into a massive mud flow that carried all before it, sweeping the surface of the limestone clear and carrying all the soil, and everything in it or growing on top, down into the valley, to deposit it on the swampy surface of the valley bottom.

The southern flank of the mountain range had a spur extending eastwards into the lowlands to east and south, and it was down the south-facing slopes of this spur that the mud flow swept. In the soil and on its surface were countless bones and teeth of animals that lived and died there, some of them natural deaths; a great many animals, however, died violent deaths from predator activity. The animals had lived in a subtropical forest, and during the long dry season the trees had lost their leaves and the plants on the forest floor had died back, so that times were hard for most of the animals living there. It was the archetypal jungle of Kipling, with extremes of wet and dry in different seasons of the year, the forest an impenetrable mass during the wet season, when both trees and ground vegetation flourished, but drying out during the dry season. Not many living animals were caught by the mud flow, however, as they had had plenty of warning, and it was only the remains of dead animals that were carried to the valley floor.

Even this event was not the end of the story, for the springs welling up through the underlying rocks started to sort and dissolve the bones, destroying many of them and mostly leaving just the more-resistant teeth. Fourteen million years later, when a road was being made through the dried-out deposits, the bones, now converted into fossils, were found, and my own story begun.

Chapter eleven

Ancient bones in the frozen continent

Jerry Hooker

The first I learnt of the possibility of a trip to Antarctica was early in 1988 when all the vertebrate palaeontologists at the Natural History Museum (NHM) were invited by the British Antarctic Survey (BAS) to compete for a place on an expedition to the frozen continent. I was lucky enough to be chosen, mainly I suspect because I don't much mind the cold and also because several of the others had only recently returned from a major expedition to torrid north Africa. Previously BAS field palaeontologists had been specialists only in animals without backbones (invertebrates), but they had recently made some finds of fossil bones of marine reptiles on an island near the Antarctic Peninsula, which caused them to seek out the collaboration of someone who was familiar with collecting and interpreting fossils of land vertebrates, or tetrapods.

The trip was scheduled for the southern summer of 1988/9 (that is, the northern hemisphere winter). It would be a short field season of no more than a couple of months, made possible because the ship we were to sail in would not just land the party and leave, but remain in the area as our base for the duration of the trip. The members of the party were to include an international group of 15 geologists including sedimentologists, vulcanologists and a glaciologist as well as palaeontologists from countries as far apart as Brazil, New Zealand and Sweden. In addition there were four biologists. The expedition was organised by Mike Thomson, Chief Geologist of BAS. The geologists were supported by one secretary, and four field assistants, who were particularly experienced in outdoor living in Antarctica. There was also a photographer to record events and, on board ship, a medical doctor.

The venue was the island group just east of the tip of the Antarctic Peninsula, named after the largest, James Ross Island. The islands lie at about 64° S latitude and are important for their rocks of Cretaceous and early Tertiary age laid down between about 120 and 30 million years ago. There are extensive exposures of these rocks because the islands get little snow in summer and hence have smaller glaciers than on the adjacent mainland, in whose snow shadow they lie.

My most important preparations for the trip involved being fitted out by BAS with the appropriate Antarctic clothing and learning to operate an all-terrain vehicle (ATV), which enables travel over rocky areas where slopes do not approach the vertical. The clothing consisted of a thick wool shirt, moleskin trousers, thick cotton sweater, synthetic wool jacket, breathable waterproof anorak and overtrousers, thermal underwear (which I never wore) and various designs of glove, sock, balaclava, boot and sun goggles. The kit was gauged according to the conditions of the latitude we would visit – and we were nowhere near the South Pole.

My own specialisation is fossil mammals and I was keen to find any tetrapods, as these were poorly known in Antarctica. I hoped they would tell us a lot more about distributions in southern continents past and present. I realised the task might be difficult, because all the sedimentary rocks in the area I was visiting were formed under the sea, not in lakes and rivers – the usual depositional sites for land animals. I was nevertheless encouraged by the recent finding of two jaws and a tooth of Eocene marsupials on Seymour Island by a US expedition and some bones of a Cretaceous dinosaur on James Ross Island by an Argentine group. The marsupials date from 45 million years ago; the dinosaur from about 80 million years ago.

Scientifically useful tetrapod fossils range from pinhead-sized mammal teeth to dinosaur bones weighing hundreds of kilograms, and the methods of collecting are therefore diverse. I needed to equip myself with a range of tools for tackling both hard and soft rocks; a sack of plaster and the various accessories for the safe jacketing of large fractured bones that would otherwise fall apart on the journey home; also a stack of large sieves of various mesh sizes and a motorised water pump for wet-sieving soft rock such as sand, the only way to recover small specimens that are sparsely distributed through the strata.

On 3 January 1989 our party of 21 flew from Brize Norton to Mount Pleasant Airport in the Falkland Islands. The flight was punctuated by a brief refuelling stop at Ascension, a tropical island composed of hummocky hills of dark weathering basalt interspersed with military installations. We were locked into a wire outdoor cage while we waited to reboard and photography was forbidden. Once in the Falklands, a battered old coach took us through a landscape of brown grass-covered hills, with grey rocky escarpments at their peaks and long moraine-like trails down their slopes. The road passed through several minefields, a legacy of the conflict seven years earlier. After an hour we reached the harbour at Stanley where we boarded the

RRS (Royal Research Ship) *John Biscoe*, which was to be our main base for the next seven weeks.

The *John Biscoe* was an ice-strengthened cargo ship with accommodation for 10 officers, 21 crew and our party. There was also a certain amount of laboratory space and a hold housing a scow (an unpowered flat-bottomed barge), a launch (usually roped to the side of the scow to power it), several Humbers (inflatables with outboard motors) and other equipment. The ATVs were stored on the aft deck when not required. There was no helicopter pad, but fore and aft decks were fitted out with gantries for lowering heavy objects over the side. I shared a small cabin with three others: Peter Bengtson, a Swedish ammonite specialist; Rob Barnes, a structural geologist; and Ian Hawes, a biologist. Double-deck bunks were economically distributed and a small porthole allowed us to keep an eye on the weather and provided a little air. The compartment where we ate our meals, drank and relaxed was called the Fiddery. I asked about the origin of the term and discovered that it was a place where Fids gather, Fids being us. Originally the term applied to members of the Falkland Islands Dependencies Survey, now the British Antarctic Survey, but despite the name change the old acronym has stuck.

In only 17 hours we had travelled from 52° N to 52° S latitude. While still in the harbour, any lingering disbelief as to our southerly position on the globe was dispelled by the sight of a group of giant petrels, a soaring black-browed albatross and distant swimming penguins. The ship got under way late in the evening. We had begun our journey across Drake Passage, reputed to be the stormiest patch of sea in the world. Consistent with several other surprises to come, the three-day crossing of Drake Passage was uncharacteristically calm. In fact, in England, I have experienced worse crossings to the Isle of Wight. We were given a guided tour of the ship including the engine room and were instructed in fire and life-boat emergency drill and use of the radios. During the trip we had access to much of the ship, including the monkey deck, which was above the bridge and afforded good views of the surrounding seascape. Many seabirds, including several kinds of albatross, petrel and shearwater, persistently followed the ship day and night and encouraged me to attempt lots of photographs. If any of the officers on the bridge sighted a whale, they would sound the hooter, whereupon there would be a rush up the narrow staircase to the deck in the hopes of a view. This way we managed to see, at least spouts of, killer, minke and humpback whales.

At the beginning of the second day at sea we crossed the Antarctic Convergence – the place in the southern oceans where the lower latitude ocean currents, in this case the Brazil Current, meet the Circum-Antarctic Current. This is accompanied by a sharp drop in temperature, which often produces mists. On the third day at sea we saw our first iceberg and that afternoon reached the South Shetland Islands with their spectacular mist-shrouded coastal peaks and glaciers. We were visited by a helicopter from HMS *Endurance*, which lifted from our deck, while airborne, a spare engine for one of the BAS Twin Otter aircraft stationed to the south at Rothera on Adelaide Island.

Around midnight we made landfall in the South Shetlands, where we picked up a biologist from Livingston Island, then sailed back up Bransfield Strait between the South Shetlands and the west side of the Antarctic Peninsula, through the Antarctic Sound and into the Weddell Sea. All the time icebergs became more numerous and ranged from small jagged ones to huge tabular bergs. The latter when on the horizon resembled distant white cities. They had broken off the ice shelf and drifted north. The colours of the icebergs varied greatly from shining white, brown, even black, to a beautiful azure blue where they showed through the crystal-clear shallow waters. On the horizon appeared a solid line of ice. As we approached this it turned out to be loose pack ice. We scrunched our way through it at a much reduced rate of knots. The ice-floes were frequently occupied by small groups of adelie and chinstrap penguins, often by crabeater seals and once by a leopard seal. We also sailed through a pod of large whales, probably fins. The weather was brilliantly sunny as we traversed this spectacular icescape.

On passing through the Antarctic Sound we left behind all this pack ice and made swift tracks for James Ross Island. In the Fiddery detailed plans were made for landing different groups of geologists and biologists on key parts of James Ross and Vega Islands. I was disappointed that there were no plans to visit Seymour Island, on which the few known Antarctic fossil mammals had been found. After finalising the plans we went up on deck to watch the sunset merge into sunrise (we were just outside the Antarctic Circle) and got our first views of the islands on which we were to land the next day. An hour later the ship unexpectedly reached the edge of a dense sheet of fast ice that encircled James Ross and Vega Islands.

The *John Biscoe* was not an ice-breaker, but tried to ram her way through the fast ice. She needed a long run to get up speed for each ram. Progress was slow and there was the danger of the ice closing back around the ship and trapping her. This operation was thus eventually abandoned. After finding that the seaways between the two islands (Sidney Herbert Sound) and between these and the Antarctic Peninsula (Prince Gustav Channel) were both blocked with ice, two of the field assistants, Tim Whitcombe and Crispin Day, skied across towards Vega Island to check the possibilities of transporting equipment across the ice, but found that significant gaps

Figure 11.1 The RRS *John Biscoe* in pack ice near
James Ross Island.

in it would make this impossible. The captain, Chris Elliot, then decided to sail eastwards in search of other possible landing places. This extent of fast ice and pack ice was unprecedented in the area for this time of year and meant that at least for now we had to abandon our carefully laid plans for landing on James Ross and Vega Islands. We found that the only landing places free of pack ice were Cockburn Island and the north end of Seymour Island. Consequently, my earlier dashed hopes of visiting the latter were suddenly resurrected.

The next day we anchored about 4 kilometres off Seymour Island, near Cape Wiman, and most of our party boarded the scow and set off for the shore. As the tide was low we had to transfer in groups to two Humbers and eventually to wade through the shallows and across the broad mudflats with all our tents, foodboxes and other equipment to the beach at Cape Wiman. As we crossed the intertidal zone, the native inhabitants came out to meet us, one actually trying to board the scow. Luckily, they were friendly as they were only inquisitive adelie penguins. Ten of us then set up camp consisting of five tents on the coastal platform.

At the end of the day, the remaining members of the party returned to the *John Biscoe*, after which they were to pay a visit to Cockburn Island. The next day we began hunting for fossils and measuring sections in the well-exposed sequence of Paleocene and Eocene soft sands and muds deposited between 60 and 40 million years ago that made up the geology of this part of the island. Each of us had our own particular areas of expertise. Martha Richter specialised in fishes, Alistair Crame in molluscs, Jim Riding and John Keating in microscopic plants, Duncan Pirrie and Mike Isaacs in sedimentology and I in mammals. We were aided by field assistants Mike Dinn and Tim Whitcombe, while Chris Gilbert made a film record of the proceedings.

The sedimentary rocks, not only of Seymour Island but also of other islands in the group, were laid down under a shallow sea between about 120 and 30 million years ago. A great thickness of strata accumulated here as vast quantities of sediment were washed down by rivers eroding the actively uprising Antarctic Peninsula. Subsidence of the sea bed prevented the sediment from filling this sea to its brim for many millions of years. Later, about 20 million years ago, during the Miocene epoch, volcanoes erupted on the peninsula and cast their lava and ash into this same sea. This volcanic outfall now forms hard rock that caps the mountains or forms steep sea cliffs.

Prospecting on Seymour Island revealed an abundance of shelly fossils at some levels, including molluscs, crabs, lobsters, sea urchins and starfish. These were accompanied by the teeth of sharks and the bones of giant penguins. The presence of nautilus shells, of leaves of the southern beech tree (*Nothofagus*) and of relatives of the

Figure 11.2 Mahogany Bluff, Vega Island, with fringing pack ice. The cliffs display Miocene volcanics of the James Ross Island Volcanic Group.

Figure 11.3 Setting off from the RRS *John Biscoe* for Seymour Island in the scow and a Humber.

monkey puzzle tree (a conifer of the family Araucariaceae) and the overall high diversity of species is indicative of conditions considerably warmer than exist in the area today. In fact, there is no clear evidence that there was much polar ice at the time (50–40 million years ago), which was the warmest interval in the past 100 million years. It was, however, difficult to envisage this ancient equable scene as we crawled in howling gales round the exposed tops of the hills where the fossils, as luck would have it, were concentrated.

I was keen to find mammals as the discovery of marsupials had been made only a little to the south. Initially the only tetrapod remains I could find belonged to penguins, but eventually my eyes alighted on an unprepossessing fragment of tooth, which I was later able to identify as belonging to a member of the extinct mammalian order Astrapotheria, previously known only from South America. It was not a

Figure 11.4 Collecting fossils in the Eocene La
Meseta Formation on Seymour Island.

Figure 11.5 Fossil leafy shoot of an araucarian
conifer (monkey puzzle relative), from the Eocene
La Meseta Formation, Seymour Island.

marsupial but the first fossil placental from Antarctica. Although only a fragment, it had an impact on theories of the colonisation of Australia by marsupials. Previously it had been thought that ice-free Antarctica between 60 and 40 million years ago, while still joined to both South America and Australia, was covered in such dense forest that only small tree-dwelling marsupials were able to travel through thereby reaching Australia. My discovery of a large ground-dwelling placental suggested more open vegetation in Antarctica at that time and the need for a different theory to explain why land-based placentals did not reach Australia until about 5 million years ago. Tim, Mike, Chris and I collected samples of the softer parts of the bone-rich strata, which contained abundant sharks' teeth. We sieved the samples in the sea in the hopes of finding more mammals. None, however, turned up there or after later examination back in England.

We had not intended spending more than a few days on Seymour Island, but events elsewhere determined otherwise. The *John Biscoe* had managed to find a gap in the pack ice and land a party on James Ross Island, but later became stranded in ice near Vega Island. Our water supply soon ran out and Mike created a temporary reservoir that we filled with icicles and pieces of iceberg to be melted as necessary. We also lost radio contact with the ship for 24 hours. This may have been due to an ionic storm, but at the time we were worried. The *John Biscoe* finally reached us 10 days after we first arrived at Cape Wiman. The experience gave the new recruits (of which I was one) an opportunity of finding out what camping in the Antarctic was all about. I had never before had the water freeze on me as I was doing the washing up. Nor had I ever been dive-bombed by a skua while sitting on the loo. Another occupational hazard was icicles in the beard. Our diet was strikingly different from the cordon bleu style we had become accustomed to on the ship. Our main food was dried and rather uninteresting, but we had individually brought our favourite herbs and spices to improve the taste of supper. I found I was needing to eat about a pound of chocolate at lunchtime every day in addition to the usual biscuits, tinned meat and fruit. Moreover, it was standard practice at breakfast to add a handful of sugar lumps (powdered with a geological hammer) and a slab of butter to the porridge. It was evident that we were all expending a great deal of energy just to keep warm. In fact, the temperature varied greatly. Sunny days without wind could be comfortably warm, but at night the sea would often freeze.

Each tent housed two people. A line of boxes with food and medical supplies between the sleeping bags formed a bench on which sat the primus stove and sundry other items. Each thick sleeping bag was enclosed in a fire-retardant blanket and rested first on a lambswool rug and then on an air bed. After an evening of cooking on the primus and with the tilley lamp burning, it was pretty cosy and not conducive to going outside to do the washing up. At this stage of the season, it was light enough inside the tent to read as late as 11 p.m. without artificial lighting. Every morning and evening, radio contact was made with the ship to ensure everyone was safe and to exchange news. We had portable solar panels to make sure the radio batteries stayed charged.

By now enough of the pack ice had dispersed to allow several selected landings on James Ross and Vega Islands. So a day later small groups boarded the scow and Humbers for their appointed destinations. Mine was St Martha Cove on James Ross Island as this was the island where the remains of the only Antarctic dinosaur had been discovered. We set up camp on the south side of the cove and explored by ATV the largely ice-free north end of the island. Our first finds were of apparently modern crabeater seal skeletons, which lay on the nearby hillsides. Some were completely enclosed in their mummified skins. Others were reduced to a scatter of corroded bones encrusted with black and orange lichens. It was interesting to see such a range of variation in their state of preservation and decay and was a reminder of the importance of the study of taphonomy (the means by which organisms become buried and fossilised) to palaeontologists when reconstructing ancient communities. We wondered how long they had lain there, whether or not they represented a single catastrophic event, and what had caused them to migrate inland up to a kilometre from the sea.

We pressed on inland to San José Pass, which divides this end of the island into east and west halves. To the north and south were ice-capped peaks with small glaciers. On sunny afternoons there would be much meltwater flowing east into St Martha Cove and west into Brandy Bay. This allowed the growth of patches of bright-green mosses that formed the only visible vegetation in this wilderness of rock and ice. From San José Pass we had a magnificent view west to the snow-covered Antarctic Peninsula. The intervening seaway was seen to be completely blocked by very large icebergs, so there was clearly still no access to the west side of the island by ship. The mountain peaks were composed of volcanic rocks of Miocene age, which resisted erosion more than the softer underlying Cretaceous sands forming the gentle lower slopes. From a distance, the latter appeared to be extensively exposed, but on closer inspection had a rather obscuring veneer of rock fragments that had eroded from the overlying volcanics. This made prospecting more difficult than expected, except in the steeper gullies, and, after a few days, my collection of tetrapods consisted only of a tooth and a few vertebrae (segments of the backbone) of extinct marine reptiles called plesiosaurs.

Soon after arriving on the island we had noticed a camp belonging to an Argentine expedition on the north side of the cove. One evening we visited them and spent a pleasant time exchanging experiences and sharing their barbecued beef and maté. The latter is an infusion made from a South American plant related to the holly, which is drunk through a metal straw with a perforated end bulb. One of our hosts, Eduardo Olivero, I soon learnt, had been the discoverer of the dinosaur bones. These belonged to an armoured type, called an ankylosaur. He was in the process of searching for more of the animal and regarded its discovery as a highly unusual event as the sedimentary rocks in which it was found had been laid down under the sea.

I felt the need to try a different approach to searching for dinosaur bones. So when Mike Thomson joined us the next day to prospect some areas west of the pass where bones had previously been found, I decided to try wet-sieving in bulk. I chose a bed of conglomerate that probably represented a shallow submarine slip in the sea bed, which I thought might contain land-derived debris, because on the surface Martha and I had found a vertebra of an extinct marine lizard called a mosasaur, a large seed and lots of wood fragments. On that evening's radio schedule there were reports of bones being found by other members of our group on Vega Island. But I decided to persist with the sieving operation before moving on.

After moving camp to Brandy Bay, Crispin and I excavated about two-thirds of a ton of unconsolidated conglomerate and towed it in trailers using the ATVs to a pond 3 kilometres away. On the journey my ATV became stuck in the mud thanks to the brilliant sunny weather that had accelerated the melting of the moraines (ice-encased debris) at the foot of the surrounding glaciers. Momentarily, my mind turned to thoughts of global warming, until we reached our destination, whereupon the weather changed and our wet-sieving operation had to be undertaken in a blizzard. We sieved the sediment to remove the fine fraction, using water pumped from the pond, then towed the concentrated residue back to St Martha Cove. We got bogged down on the way several times so that the six-kilometre journey took three hours. Luckily at that stage I didn't know the samples would yield little of interest. That evening, the news over the radio that more bones had been found on Vega Island encouraged me to return to the ship as soon as possible and make for that place.

Early next morning, I spent a hurried hour and a half on the ship unloading and preparing for the next foray. I left by Humber and exchanged places with Peter Bengtson at the Cape Lamb camp on Vega Island. I quickly looked at some of the bones that Tim had found (they belonged to a plesiosaur) and made straight for Sandwich Bluff, the peak that rose behind our camp. There, in a similar landscape to that of James Ross Island, but with better exposed Cretaceous strata, Tim showed me

Figure 11.6 Wet-screening operations on James Ross Island.

the site of his finds. There was a scatter of bones partly obscured by snow. I began recording the position of each bone prior to collection to help me restore them to their correct positions in the skeleton. This was a very bleak place. The strong wind made it the coldest day of the trip yet and my shivering made the recording task difficult.

The sunshine hastened the snow melt and the plesiosaur soon became fully exposed. After collecting all the visible bones, Tim and I dry-sieved the surface sand in the vicinity to avoid leaving anything important behind. When pieced together the specimen consisted of 29 vertebrae and a couple of paddle bones, but unfortunately nothing of the skull. The next morning I looked at the rest of the bones that had been found before my arrival at Cape Lamb and, among those that Peter Bengtson had collected, I immediately recognised with amazement several vertebrae that clearly belonged to a dinosaur. Others found at the same spot appeared to be parts of limb bones from the same animal. Tim had been with Peter when he found the bones and we tried to locate the spot, which was on the low slopes behind the beach about a kilometre from the camp. Peter had built a cairn there but Tim had unfortunately dismantled it as he thought it was no longer needed and quite rightly did not wish to perpetuate artificial structures in the Antarctic. We hunted around but found no more bones.

That evening I spoke to Peter by radio and he gave me a more precise location for the bones with respect to a distinctive sandstone layer that we had both noticed. So the next morning I set off again for the site with greater hope of finding more of the skeleton. After only a few minutes I found the jaws of the dinosaur – all four of them, upper and lower. They were broken into several pieces but the breaks were sharp and they fitted back together well. The active teeth had nearly all been worn away through weathering on the surface of the ground, but the sides of the jaws were broken in such a way as to expose several unerupted teeth (dinosaurs, like crocodiles, replaced their teeth continuously through life, unlike us humans who have only two sets, milk and permanent). The teeth were leaf-shaped with vertical ridges and so I knew that it was a plant-eater and one of the bird-hipped types (Ornithischia). I was greatly spurred on by this discovery and spent all day picking up every piece of bone I could find in the area, which covered about 150 square metres. This was a very

Figure 11.7 One of the lower jaws of the dinosaur soon after its discovery on Vega Island.

absorbing task, aided by the peaceful nature of the setting. The only sound was the distant crackling of icebergs as they lazily jostled for position with the ebbing and flowing of the tide. Some of the bones were partly enclosed by limestone and these were the more complete ones. Others had been shattered into pieces by frost. This process could have been operating for many years as it was impossible to know low long ago the skeleton had been exhumed from its ancient grave by the agents of erosion.

Martha, Alistair and Duncan, who had spent the day on Sandwich Bluff, had been successful in finding more plesiosaur bones and other important fossils. We were all rather euphoric and celebrated the dinosaur discovery with a barbecue (or the closest we could get to a barbecue given our resources). The new dinosaur was clearly different from, and much more complete than, the ankylosaur and it was obviously important to recover as much of it as possible. As it all appeared to have weathered out onto the surface, the best method seemed to be dry-sieving the site. Unfortunately we were scheduled to leave the next day (7 February) so that several of our party could rendezvous with HMS *Endurance* and be landed in more remote locations by helicopter. At the eleventh hour we heard on the radio that HMS *Endurance* had been damaged by an iceberg and would be delayed while she received first aid at Deception Island in the South Shetland Islands. Our departure was therefore put off for two days, which enabled me, with Tim's and Duncan's help, to sieve the surface sand over a considerable area of the hillside around the dinosaur site. I ended up with thousands of bone fragments and was faced with having to do the biggest jigsaw puzzle of my life.

It was strange to see these fossil bones of the dinosaur surrounded by the remains of its contemporaries – ammonites, nautilus and extinct fan-mussel shells (*Pinna*) – that had lived in such a different environment, the open sea. Such finds of land-derived dinosaurs in marine strata are indeed rare although some classic examples have long been known from Britain: for example, *Cetiosaurus* from the Oxford Clay and the Maidstone *Iguanodon* from the Lower Greensand. Their remains are believed to have drifted out to sea after they had died. Decomposition of large vertebrate carcasses in water normally includes a long period of floating during which bones drop from the body and become scattered over a wide area of sea floor. Therefore, to have remained as an articulated skeleton on its journey to the bottom of the sea, a vertebrate carcass may have partly decomposed and dried out on land first, the bones then being held together by more resistant ligaments and tendons.

Back on ship there was a party that evening to celebrate the end of our stay in the James Ross Island area. The next day the captain found that our way east out

Figure 11.9 Dry-screening at the dinosaur site, Vega Island.

of Sidney Herbert Sound was blocked by ice and we had to stay put for now. Helicopters came to take several of our team to HMS *Endurance*, which had now limped as far as Hope Bay. I then began the dinosaur jigsaw puzzle helped by Jane Francis, Crispin Day and Gill Clarke. Almost the entire ship's crew and complement of scientists came to see if they could fit a bit of bone onto one of the main pieces, and to photograph the proceedings. It was pleasing to find so much interest shown in just one fossil.

After nearly three days of cruising around, the *John Biscoe* eventually found a passage north between James Ross and Vega Islands. We then sailed north-east to Hope Bay to rendezvous with HMS *Endurance*. From there we were to move east to Signy Island in the South Orkney Islands before returning to the Falkland Islands. This leg of the journey was to land one of the biologists, Paul Brazier, for two seasons, and to unload vital equipment and supplies for the base there. The detour at first gave me a chance to continue the jigsaw. After a while, however, rolling seas made this hazardous and I had to pack the specimen safely away.

Figure 11.8 (*Opposite*) The camp at Camp Lamb, Vega Island.

Although the jigsaw was far from being completed, in addition to the four jaws, including small parts of the skull, I now had 15 neck and trunk vertebrae, both humeri (upper arm bones) nearly complete, most of the scapulae and coracoids (shoulder girdle), part of a lower arm bone, a couple of pieces of pelvis and many rib fragments. Even though the skeleton was lacking the tail and nearly all the hind quarters, it was clearly very different from the earlier found ankylosaur. It was later identified back at the NHM by dinosaur specialist Angela Milner as a new genus of primitive iguanodontian, one of the bipedal ornithopods. The individual was estimated to measure about 4 or 5 metres in length. This is the first Antarctic ornithopod and, together with several related types from southern Australia, provides evidence of a diversification of small bipedal plant-eaters in the ancient south polar supercontinent called Gondwana. By the time this dinosaur was living (about 75 million years ago), the supercontinent had begun to fragment and drift apart into the component parts that we know today as South America, Antarctica, Australia, Africa and India. Also at this time, the climate was temperate and the area was clothed in vegetation, making it a very different place from today.

We passed Elephant and Clarence Islands and a day later the triple spikes of the Inaccessible Islands came into view to starboard with distant Coronation Island of the South Orkneys to port. We had followed an arcuate course, keeping well to the north in an attempt to avoid the pack ice that was still drifting out of the Weddell Sea. As we approached Coronation Island pack ice appeared, but we made it through and sailed down the spectacular narrow strait between Coronation and Signy Islands to the base. We spent a day unloading 32 tons of cargo. Our reward was to be allowed to walk round Signy Island to view the wildlife. Signy is at a slightly lower latitude (61° S) than the Antarctic Peninsula and there is more vegetation. In addition to the extensive moss banks we were able to see the only two species of flowering plants that exist on the Antarctic Continent: a grass and a pearlwort. We visited a chinstrap penguin colony and also had close encounters (too close from an olfactory point of view) with fur seals and elephant seals.

The return to the Falklands took three days and was uneventful. I spent the time packing up fossils, writing reports and watching seabirds and whales. By 20 February we were back to civilisation – quite a shock – and were ready to fly home. The trip proved more successful for me than I could ever have imagined. This success was, ironically, largely thanks to unexpected events that altered the original plans for the expedition. If it had not been for the exceptional northern extent of the pack ice we would never have landed on Seymour Island and if HMS *Endurance* had not had a close encounter with an iceberg I would not have had time to make a thorough job of collecting the dinosaur. The weather on this trip had been exceptionally good overall and we had not been tent-bound for a single day. Such is the unpredictability of doing fieldwork in the frozen continent. All members of the group had had important successes, whether it was finding new key fossils, understanding how both the sedimentary and volcanic rocks had formed, or investigating how the climate had changed through the millions of years using fossil tree rings or isotopic signals in fossil shells.

I want to express my thanks to the British Antarctic Survey and the Transantarctic Association for funding me on this expedition and to all my friends and colleagues who were with me on the *John Biscoe* and in the field for their help and comradeship.

Chapter twelve

Arabia Felix: fossilised fruits and the price of frogs

Peter J. Whybrow

At the busy time of the commercial day we found the principal Arab citizens reclining on divans chewing khat: *later he took me to an Arab café where the lower class congregate; here, too, was the same decent respect for leisure; the patrons reclined round the walls in a gentle stupor, chewing* khat.

Remote People, Evelyn Waugh, 1931, on Aden

The afternoon grid-lock in Sana'a seemed worse than usual. In addition, the streets were unusually full of milling Yemenis acting like crowds who have just seen their national team lose a crucial football match. Ian, who was driving, thought that a national celebration was taking place – but the mood of the throng lacked any jubilation. Suddenly, a man commanded us to leave the area immediately and drive down a side street. We detoured along the rubbish-strewn back-roads passing by dour muddy-brown buildings that all looked the same. By dead-reckoning – the sun was just visible through the permanent dust haze – we returned safely to our lodgings in the American Institute for Yemeni Studies (AIYS).

The shooting started soon after and we sped upwards through the buildings' six storeys to the roof to see if yet another Yemen revolution was beginning. Long bursts from automatic weapons sounded off, troops advanced to somewhere in skirmishing order and wisps of tear gas started to come in our direction – thankfully the only part of the mini-battle that did. Unseen to us, but like a river changing direction numerous times while in flood, occasional gunfire erupted from different areas near the souk until a crowd with a prime view of the combatants from a bridge over a wadi suddenly, like a wave, about-turned and ran down side streets. At this point the mini-battle waned and for a few hours life in Sana'a was abnormal – it was quiet. The next day Yemeni colleagues told us that the ruckus started when a policeman tried to admonish a military commander for driving through a red traffic light. The commander shot the policeman, probably because of a long-standing family blood-feud, but then the police and military decided to have a shoot-out. Coupled with this were undertones remaining from the unification in 1990 of the northern Yemen Arab Republic with the southern Soviet-backed People's Democratic Republic of Yemen (PDYR) – the former Aden Protectorate.

The merger of the two Yemens was one reason for our 1991 visit because the former PDYR had become more accessible to visitors, especially after the Russians and East Germans had departed. The palaeontological reason was that I had been telling colleagues of my idea, published in 1983, that some mammals migrating out of Africa between 30 and 5 million years ago might have used a geologically ancient landbridge to cross into Yemen from Ethiopia – the African/Arabian connection.

Examine any map of the southern part of the Red Sea and it can be seen that a slight clockwise rotation of Arabia would close the Red Sea and the Gulf of Aden. The southwestern toe of Arabia would easily fit, like a jigsaw, into the area between the Ethiopian and Somali plateaus known as the Afar (or Danakil) Depression – the junction of three of the world's greatest rift systems – the East African, the Gulf of Aden and Red Sea rifts that now form the longest continuous faulted feature known on the Earth's continents.

Within the African rift system, in Kenya, Uganda, Tanzania and Ethiopia, fossilised bones of, for example, ancient elephants, large cats and other carnivores, primitive horses, giraffes and hominoids, have been found since the 1920s from rocks of Miocene age – 20–5 million years ago. Some of these animal fossils are very similar to those in rocks of similar ages in southwestern Asia and, with the Red Sea open to the Mediterranean during the Miocene, the only terrestrial route from Africa to Asia might have been via Yemen. The Red Sea eventually opened to the Indian Ocean about 5 million years ago; the marine gateway is still very narrow. The area where the Gulf of Aden becomes the Red Sea lies off southwestern Arabia, the Strait of Bab el-Mandeb – the Gate of Tears. Within the Strait, about 20 kilometres from the eastern coast of Africa lies the island of Perim and just 5 kilometres to its west is Arabia. Southern Yemen therefore seemed to be a good place to seek fossil evidence to test my idea.

Safely departing Frankfurt and hoping we would similarly depart Cairo, stories of Yemenia not paying airport landing fees were rife, Ian Tattersall and Jim Clark, both from the American Museum of Natural History in New York, and myself arrived at Sana'a airport in Yemenia's antique plane – one of those Boeings where water anoints your head during landing. We were, at 2 a.m., surprisingly greeted by Hamad el-Nakhal, a Palestinian (now back in Gaza) who taught geology at Sana'a University. Hamad drove us through deserted streets and languidly policed check points – Sana'a had a night curfew – to our lodgings at the AIYS, where Ian had stayed during his 1988 visit to 'northern' Yemen. The multistoreyed AIYS building is one of many architectural splendours in Yemen. It has thick stone walls, at least a metre thick for the lower floors, and a steeply ascending winding staircase with 50-centimetre-high stone treads. The upper rooms feature mud-brick walls and high, palm-wood beamed

Figure 12.1 Architecture in Sana'a.

ceilings – a design based on centuries of experience on how to keep occupants cool during the hot Yemeni summer.

The next day we visited the Ministry of Oil and Petroleum Resources to present our salaams and to find out what logistic support might be provided for our research. Ian, normally a palaeoprimatologist (somebody who researches fossil monkeys and apes), had obtained a grant from the National Geographic Society to look for dinosaur bones. Word of a previous, unpublished, discovery of such bones in Yemen by oil company geologists had got about. Jim, who had just returned from dinosaur hunting in Mongolia, was armed with heavy hammers, chisels and other dinosaur extraction impedimenta to dig the beast out of the rock – should we find it. That was our first objective. Our second – and to myself and Ian the more important one – was to visit southern Yemen to discover fossilised mammals. I also wanted to meet with British Petroleum geologists in Aden so as to set up a visit for two colleagues wanting to study the Jurassic marine rocks in the Hadramaut.

These objectives were explained to the Yemeni geologists, who for many years had been searching for geological material of economic importance. Consequently it took them a few days to grasp our esoteric and, to them perhaps, bizarre needs. For any geologist, however, several days away from a desk in a cacophonous, dusty city, built at 3000 metres in an old volcanic crater, to find fossils 'out in the country' would be like a holiday. To our Yemeni colleagues, participation in our work, as we were to find out, was to become more than a holiday, especially for those who had never before visited the south.

Perhaps as a test of how ourselves and the Yemenis would get on together, a three-hour trip to a site found by Gemil Saif – one of Hamad's students – was arranged. The site, it was said, had produced the remains of fossil frogs. These, of course, are neither dinosaurs nor mammals but at least we would escape from Sana'a, see some of Yemen and possibly find other animals preserved with the amphibians. With ourselves and Hamad in the borrowed AIYS Toyota four-wheel drive, and several men from the ministry crammed into their own Toyota, we arrived at the site having passed spectacular mountainous scenery, several hill-top forts, stunning small-town multi-storeyed houses and nearer to the site, in country formed by volcanic rocks, tiny stone-walled fields in which a few sticks of maize barely survived. An embankment by one of the fields exposed slate-like rocks that we attacked with hammers and chisels.

After several hours of hacking and hewing and splitting slabs, we found just four poorly fossilised frogs. Oh well, but at least we got to know the Yemenis one of whom, Mustafa Latif As-Saruri, was to become a good friend. He was from the former People's Democratic Republic of Yemen and had received most of his geological

Figure 12.2 Hill villages amongst the volcanics.

training from the East Germans. Passionate about geology he had been studying for numerous East German degrees for what seemed like decades and had travelled throughout most of southern Yemen, especially to the politically difficult area close to the border with Oman. Mustafa was to be our scientific and guiding link to the palaeontological potential of southern Arabia.

Our apparent excitement on finding the frogs – the first from Arabia – prompted the first of several misunderstandings. We were anxious to find the dinosaur site for Jim, who planned to leave in 15 days, and we wanted to examine the Jurassic rocks as soon as possible. Our colleagues, however, exuberantly told us that we were to be taken to yet another frog-bearing site, this time some four hours drive from Sana'a. We duly arrived in yet another large valley surrounded by black volcanic rocks. Most of Yemen consists of volcanic material, ash and basalts, produced during Arabia's split from Africa, a process still continuing at the rate of about 2 centimetres a year. Intense geological activity associated with the splitting, from about 40 million years ago, has continued almost without interruption to the present day. The volcanic Mount Haras erupted in 1937 and around Ma'bar the 1982 earthquake claimed 2000 lives.

We headed for a long ridge of steeply dipping white-coloured rock. The surrounding hills exposed alternating layers of a similar rock known from work in the 1960s by French geologists to be sediments deposited in ephemeral freshwater lakes

Figure 12.3 (*Opposite*) The site for fossil frogs.

formed during brief respites of volcanic activity. On hands and knees we peered at the rock for some time until the first frog was found. About 2.5 centimetres in length, only the shape of where its bones had been – skull, ribs and limbs – were preserved. Sad-looking fossils of amphibians that died, we supposed, when their lake had become either too hot for survival or had been engulfed with volcanic ash. Oddly, most of the frog fossils were orientated in the same direction; had they attempted to escape from the catastrophic eruption? Many more were seen on the slope but most could not be removed without mechanical equipment. During the time it took us to hammer out about eight specimens, some locals had arrived and squatted beside us but offered no help. Mustafa, who had been talking to a local some distance from our frog site, wandered over.

'That Man,' he said to me in a whispered, embarrassed and somewhat frightened tone, 'has an AK47 and wants your vehicle, your passports and all your money. You are on his land.'

'Land,' I said, 'this isn't land, its rock and rubble without plants. There aren't even goats around here!'

'No matter, you are on his property, you are digging it up and removing it and because you are Europeans what you are digging up must be valuable.'

'Well, he seems to have started at his maximum negotiating position!' 'Ian,' I shouted, 'come over here; how much money did you bring?'

Ian, who had been forcibly removed by the Yemen military from a campsite near the Saudi border in 1988, mouthed several curses at the AK47-wielding extortionist. Mustafa negotiated while staring at the ground and prodded the slabs containing the frogs with his boot, presumably hoping that there might be an amphibian resurrection and the problem would jump away. The Man negotiated while staring at the hills and lovingly fingered his weapon – did he have reinforcements?

'OK, he will accept 1000 Riyals,' Mustafa informed Ian, who, normally a theatrical character, seemed subdued, pulled out his wallet and said 'Let's get out of this looney-bin, we just spent $70 on excavating **** frogs.'

The next two days were spent trying to find the dinosaur site, but it remained elusive. It was time to go to Aden.

The distance from Sana'a to Ta'izz is about 250 kilometres along the high volcanic plateau. Occasional monsoon rains irrigate the volcanic soils. Yemen was once part of an area known as Arabia Felix; happy Arabia, an area of fertility in a region otherwise known for its vast sand deserts and extremes of temperature. Unfortunately, and for a minimum of 50 years, economically sustainable produce farming in Yemen has been almost non-existent and the prime cash-crop has been *qat*. This shrub has

Figure 12.4 Jim collecting frogs; Mustafa (left) and a local offer help.

leaves like bay that when chewed for some hours have a soporific effect on the chewer. *Qat* parties dominate male Yemeni life in the afternoons and, like fine wines from certain European vineyards, there are regions of Yemen that produce the best and most expensive *qat* and we were approaching one.

The road from the high plateau suddenly and steeply descended. We passed the remains of a tank dating from one of Yemen's many civil wars. Inexplicably it was not only pointing up the almost 1:1 gradient but also its barrel was flaccidly bent. The lead Yemeni Toyota slowed considerably; worried about their brakes, we thought, or trying to find a fossil locality? Another misunderstanding; no, they were checking out the *qat* fields. Ian at this late afternoon hour was concerned about us finding any accommodation in Aden. But for the Yemenis from Sana'a this *qat* jaunt was a holiday, like having a choice from the sweet factory rather than having to buy what was on offer in the sweet shop. While Ian fumed, I took photographs of the magnificent succulent plants that grew in this arid area. These, apart from Mocha coffee and frankincense, are some of Yemen's most unusual endemic botanical assets. After endless negotiations with the *qat* field entrepreneur our colleagues wandered aimlessly (had they had tasters of the crop?) back to their vehicle clutching bundles of

Figure 12.5 The *qat* fields in southern Yemen.

leafy twigs that must have cost them much more than the price we paid for our fossil frogs.

We arrived in Aden at about 7.30 p.m. I had mentioned the name of the hotel where the British Petroleum geologists were based and we were first taken to the splendid Swiss based Movenpick Aden Hotel – but it was full. Another sea-front hotel was also full and a problem especial to visitors surfaced. A foreigner in a Yemen hotel has to pay in hard currency of which we had little, Yemenis pay in local currency at a much reduced rate. We waited in the vehicle while Mustafa went into a rundown place at the back of Aden's old waterfront and returned to say that several rooms had now been booked for us. We trooped in and, while eyeing the notice asking us to deposit our guns with the receptionist, we suffered the vituperative wrath of the manager who, seeing foreigners, realised he had been duped out of a lucrative hard-currency profit. As we had found in Sana'a, the black-market rates were very advantageous.

Figure 12.6 One of Yemen's magnificent endemic plants – almost 2 metres high.

An exceedingly long visit took place the next day at Mustafa's office at the former PDYR oil ministry. We were taken from room to room to be introduced to senior bureaucrats, all of whom sat in similar fake leather chairs. Conversation languished – we were anxious to have a look at Aden – but these chaps were worried and they hesitated to find the right words. Their communist backers had departed, north and south Yemen had started a shaky, unconsummated honeymoon, the south was hoping for major oil discoveries but the north had the backing of Islamic clout from Saudi Arabia. Also worried were many Yemeni women, who, since the communist National Liberation Front takeover in 1967, had achieved what few Arabian women could ever contemplate – a career. We agreed on a plan. The 'north's' vehicle would return to Sana'a, loaded with *qat* no doubt, and we would leave in two days with the 'south's' vehicle for Mukalla, where camping had been arranged. Meantime, British Petroleum had invited us to the British Embassy that evening.

The Embassy, whose personnel were soon to move to Sana'a, was located in an isolated area of Aden and behind 4-metre-high walls surmounted by numerous coils of nasty razor wire. The gate-guard barely nodded to us. We passed through rooms displaying, amongst other events, notices of tennis matches, where recent well-thumbed copies of the *Daily Telegraph* and the *Daily Mirror* were scattered in lounge chairs, and where one door beckoned us to the Club. The Club was heaving. Most of the oil company staff in Yemen seemed to be there and it was as full as any pub in London's King's Road on a Saturday night. Canadian Occidental staff dominated the small room and seemed to have been celebrating for some time yet another of their oil strikes. In contrast, our British Petroleum colleagues were subdued as their concession near the Saudi border did not appear promising. I finished my palaeontological business with them and we left for a nocturnal walk, after having been warned not to go to a certain beach where human remains, probably from the 1986 war, were being exhumed and re-buried by the Indian Ocean.

As a front seat passenger in the Yemeni government Toyota I experienced a unique style of driving previously encountered only along the Grand Trunk Road in Pakistan. This is to drive the 500 kilometres to Mukalla as fast as possible, avoiding potholes, donkeys and sheep, overtaking smaller vehicles on sinuous bends and people by deft flicks of the steering wheel *à la* Damon Hill. This hair-raising experience ceased as we stopped at a point where the coast road veered north to ascend into the mountains. We had stopped to climb a volcano – a dormant, geologically recent one. Since leaving New York/London Ian, Jim and especially myself, had experienced little physical effort and we were now to climb an almost vertical 500-metre-high pile of fine-grained ash in a temperature not less than 38 °C.

Figure 12.7 On the rim of the volcano.

Mustafa bounded to the top assisted by his mountain-man-like gait and by his fitness from years of measuring rock thicknesses on near vertical cliffs. We, on the other hand, zigzagged upwards in a cloud of volcanic dust, pausing only to gasp that this was not one of our planned objectives. Breasting the rim of this unfossiliferous pile of ash with my leg muscles dancing to the tune of St. Vitus, I collapsed into the dirt. But my soul was rewarded by the spectacular scene of a deep-blue, almost circular pond of sea-water in the crater's floor that contrasted with its black volcanic slopes and, amazingly, the deep-green foliage of mangroves that were just surviving on the crater's beach. Mustafa had bounded down the rim to see them but we, mindful of the heat and of the time, turned about and stumbled downward to our vehicle.

On a dusty hill on the outskirts of Mukalla, we were deposited at our campsite. Hitherto, we had imagined a site under palms, cooking over a fire and wondering at the clarity of the star-lit night. Well, we had the starry night but had to bed down in one of a hundred plywood-clad bungalows recently vacated by, it was said, Russian gold miners. Bits of this temporary part of Mother Russia littered a rapidly decaying and surreal place. Cyrillic signs pointed to somewhere; termite-chewed pictures of snow-clad peaks were caught in thorn bushes along with discarded Dolichnaya bottles; a locked projection booth at the open-air cinema still stored cans of film and a projector; and sad abandoned cats slouched around the still-functioning camp restaurant, where we ate what I thought was the best meal we'd had so far in Yemen – a mutton curry with heaps of spiced rice, fresh fruit and copious amounts of sweet Arab tea.

Since the Middle Ages, Al Mukalla had been the jewel of this part of the Arabian coast. It was the main port for the Hadramawt kings at Shabwah, who once controlled part of the international trade that exported frankincense to perfume the ancient temples in Egypt, India, Greece and Rome. Built on a striplet of land between the sea and a startling backdrop of mountainous crags it is now merely a fishing port. With the stench from piscatorial offal and waves of mind-curdling heat wrecking our olfactory and cerebral sensibilities, we walked along a grubby creation of modern town planning – the sea front had been filled to reclaim the shallow harbour to, we were told, create more space for constructing suburbs. While having a fish biriani in the town's 'best' restaurant accompanied by flies and Mukalla's locals noisily being phlegmatic, we pondered on the destruction of a once remarkable view from the now buried foreshore, arguably Mukalla's greatest tourist asset – white-painted, multi-storeyed buildings in front of steep mountains barring easy access to Yemen's mysterious interior.

Mustafa guided us to the area where he thought we might find vertebrate fossils. Here we noted two facts of southern Arabian geology that would make our search difficult. In contrast with Europe or northern America there is little soil concealing the rocks whose fossilised story we were attempting to unravel. This fact was, after all, one reason for our visit to Yemen; no rain equals no soil or vegetation to conceal potentially fossiliferous rocks. No rain, however, means no erosion by water and all slopes and valleys were full of rock rubble produced by temperature extremes that had blasted off pieces of rock over tens of thousands of years – like a stone thrown into a fire suddenly exploding. Second, and unlike the almost flat, unfolded succession of rocks from which I had collected Miocene vertebrates in eastern Arabia, in Yemen the splitting apart of Arabia from Africa resulted in numerous fractures of the Earths' crust – faulting. Such faults, called grabens, produce a staircase-like geological succession with each several kilometre thick 'tread' showing the same bits of old and young rocks that had progressively slipped into the widening Gulf of Aden. We

Figure 12.8 High Street Mukalla

decided that the moonscape we were in would not easily yield fossils as several years of dedicated mapping would be needed just to locate rocks of the age that interested us. Mustafa suggested, therefore, that we should go further inland and approach the coastal mountain belt from its northern side.

Along the winding, dirt road up Wadi Hajr I saw vertical cliffs of Jurassic marine limestones that looked superficially like the cliffs at Lyme Regis in Dorset and I knew that my Natural History Museum palaeontology colleagues would have enormous fun during their forthcoming visit with British Petroleum geologists. After many stops when our Yemeni friends perused road-side stalls (no *qat* here; the only things they ever bought were water melons),we arrived late at night at Al Gool. A pack of baying dogs enlivened the night during our wait for someone to be woken up to find the key to the government accommodation.

We had stayed at the Russian Gold Miners Camp and we were now to experience the Chinese Rest House. Eight rooms led off each side of a courtyard in which a few blades of a grass-like plant mingled with fall-out zones of melon seeds that

Figure 12.9 Deeper into Wadi Hajr.

had once been skilfully projected outwards from each door to the 16 rooms. Rooms lacked electricity, beds or bedding; rooms with beds lacked electricity, rooms with electricity lacked beds, and so on. We humped beds and baggage to what we thought was the best room. The light worked but the ceiling-mounted fan threatened to unwind and fall from its support. We then discovered that the rooms without lights but with beds were those where the mosquito screen had not been broken. Our next concern was food. Anticipating camping in tents, we had purchased numerous tinned goods in both Aden and Mukulla and it was now time for Ian to demonstrate to our colleagues the best of European cooking. The sight of a strange sludge of onions, tomato paste, tinned Arabian beef-like sausages and rice was met with hungry horror by all except Ian. Tea was rapidly brewed by the locals who consumed gallons of it, perhaps to aid their digestion.

The tortuous road to the next area that might hold promise of vertebrates was no more than a narrow camel path followed occasionally by the vehicles' left or right side wheels. We bumped and jostled along the cobble-strewn wadi and meandered up to the watershed spying freshwater fish and gastropods in reed-lined pools. Parked near a hilltop town under a rocky overhang that gave some shade, we spread out to search part of an area that Mustafa had mapped a few years before. At that time he had noted continental rocks, or at least estuarine sandstones, and found some petrified wood. We plodded around in the furnace-like heat, dodging under the speckled shade of thorn trees and, after the first painful contact, not picking up seemingly red hot rocks that had been baking since dawn in the full glare of the sun.

Littering a flat area between almost-vertical beds of yellow-red sandstones I saw rust- coloured shapes recognisable as poorly preserved twigs and branches. They were not silicified, a type of preservation that might have given clues about the type of plant and its environment, but were just sandstone casts. On hands and knees I, with difficulty, crawled over this area and soon found reddish-black shapes similar to date stones. Both their weight and their hardness proved that they were fossils and they seemed to be plant seeds. Later, soon after I found my camera had ceased functioning, I found two tomato-size lumps and several other pieces of plant. Where their outer 'skin' had been fractured numerous seeds could be seen preserved in-place and these identified the 'tomatoes' as fossil fruits. This was good news as their discovery indicated that land was nearby when the plants were alive and we might at last find some fossil vertebrates. But what was their geological age? From Mustafa's geological knowledge, and from the marine fossils we had seen in nearby limestones, we concluded that the plant fossils were from the middle part of the Eocene epoch, probably about 45 million years ago. This age was some 20 million years older than we had hoped for. From a distance Jim shouted that he was excavating bone from the side of a small hill. I plodded over to see the rib of a marine mammal, probably a sea cow.

It was time to go. Back in Sana'a, in a restaurant located on the ground floor of Sana'a's former up-market brothel, Ian, Jim and myself debated our experiences and future plans over a modern expression of the Arabian/African link – an Ethiopian

Figure 12.10 Village near the fossil fruit site in Wadi Hajr.

meal – comprising a dark-coloured extremely hot sauce with chopped meat and bones served on something (I hesitate to call it bread) very much like polyurethane foam. We hoped, with Mustafa, to examine the region closer to the Oman border. This would involve much planning and we would really have to camp. There are no socialistic camps or rest houses in the area that even Mustafa said was difficult country – too much heat and no water. Plans, of course, are controlled by events. Later, the unconsummated marriage of southern Yemen with the north ended with war that resulted in Saudi Arabia having greater hegemony over south-western Arabia. One of Mustafa's son's had an eye injury from shrapnel during the battle for Aden and almost all of the 'south's' vehicles used for geological work were wiped out.

Mustafa's undiluted passion for geology has not withered and he himself seems to have professionally survived the war, for he is now the Chairman of the Yemen Stratigraphic Commission. Mustafa, myself and Margaret Collinson of Royal Holloway College, University of London, one of the UK's few palaeobotanists, have published a preliminary account of the fruits, seeds and local geology. The Yemen fossil flora is similar to specimens found in Egypt. The continental climate in the Middle East at that time was perhaps tropical and, in Yemen, woody plants and freshwater aquatic water lilies thrived near to a tropical sea. Before the millennium is out, a return is planned to Arabia Felix, to continue examining its magnificent but complex geology and to add more fossils to its burgeoning palaeontological heritage.

Afterword

Much of this book is about experiences of British palaeontologists in foreign countries. It would be nice to have had a chapter in the same style about the recovery of fossils in Britain, written by someone from, say, Africa or Asia who would indicate how supremely ridiculous are the British.

Anonymous review of the manuscript for this book, 1998.